Exercise and Cancer Recovery

Carole M. Schneider, PhD
Rocky Mountain Cancer Rehabilitation Institute
University of Northern Colorado

Carolyn A. Dennehy, PhD
Professor Emeritus, University of Northern Colorado
Second Wind Cancer Rehabilitation

Susan D. Carter, MD
Rocky Mountain Cancer Rehabilitation Institute
University of Northern Colorado
Regional Breast Center of Northern Colorado
North Colorado Sports Medicine

Human Kinetics

Library of Congress Cataloging-in-Publication Data

Schneider, Carole M.
    Exercise and cancer recovery / Carole M. Schneider, Carolyn A.
Dennehy, Susan D. Carter.
        p. ; cm.
Includes bibliographical references.
    ISBN 0-7360-3645-8 (hard cover)
    1. Cancer--Exercise therapy. 2. Cancer--Physical therapy. 3.
Cancer--Patients--Rehabilitation.
    [DNLM: 1. Exercise Therapy. 2. Neoplasms--rehabilitation. QZ 266
S358e 2003] I. Schneider, Carole M., 1949- II. Dennehy, Carolyn A., 1949-
III. Carter, Susan D., 1957- IV. Title.
    RC271.P44S35 2003
    616.99'4062--dc21

                                2002156702

ISBN: 0-7360-3645-8

Acquisitions Editor: Michael S. Bahrke, PhD; **Developmental Editor:** Elaine H. Mustain; **Assistant Editors:** Maggie Schwarzentraub and Sandra Merz Bott; **Copyeditor:** Joyce Sexton; **Proofreader:** Jim Burns; **Indexer:** Sharon Duffy; **Permission Manager:** Dalene Reeder; **Graphic Designer:** Fred Starbird; **Graphic Artist:** Dawn Sills; **Photo Managers:** Les Woodrum, Kareema McKlendon; **Photographer:** Scott Drum; **Cover Designer:** Robert Reuther; **Art Manager:** Kelly Hendren; **Illustrator:** Craig Newsom; **Printer:** Sheridan Books

Printed in the United States of America   10   9   8   7   6   5   4   3   2   1

Human Kinetics
Web site: www.HumanKinetics.com

United States: Human Kinetics
P.O. Box 5076
Champaign, IL 61825-5076
800-747-4457
e-mail: humank@hkusa.com

Canada: Human Kinetics
475 Devonshire Road Unit 100
Windsor, ON N8Y 2L5
800-465-7301 (in Canada only)
e-mail: orders@hkcanada.com

Europe: Human Kinetics
107 Bradford Road
Stanningley
Leeds LS28 6AT, United Kingdom
+44 (0) 113 255 5665
e-mail: hk@hkeurope.com

Australia: Human Kinetics
57A Price Avenue
Lower Mitcham, South Australia 5062
08 8277 1555
e-mail: liahka@senet.com.au

New Zealand: Human Kinetics
P.O. Box 105-231, Auckland Central
09-523-3462
e-mail: hkp@ihug.co.nz

To our cancer patients at the Rocky Mountain
Cancer Rehabilitation Institute and all cancer survivors.

Contents

Preface

Cancer and the side effects of cancer treatments leave patients struggling to regain the quality of life they experienced before cancer diagnosis. The majority of patients will experience an array of symptoms such as nausea, pain, fatigue, and depression, to name just a few.

In 1998, the Fatigue Coalition conducted a survey on 379 cancer patients who had been treated for breast, prostate, lung, or skin cancer or for leukemia or lymphoma. Seventy-six percent of these patients experienced fatigue. Of the 301 patients who experienced fatigue, 60% felt that fatigue affected their lives, outweighing nausea (22%), depression (10%), and pain (6%). Economically, treatment-related negative side effects are a major concern. Seventy-one percent of employed patients missed one or more days of work a month; 31% missed a week per month; and 28% related that they were forced to quit work because of cancer treatment-related symptoms. Results of this survey, similar to those from an earlier survey by the Fatigue Coalition, showed that 59% of patients had difficulty socializing with family or friends, 37% had problems maintaining interpersonal relationships, and 30% had difficulty with intimacy. The majority of patients had problems with daily activities such as cleaning the house (69%), running errands (56%), climbing stairs (56%), and walking distances (69%). It has been observed that cancer patients have less vigor and are more fatigued than patients who have had myocardial infarctions. Cancer treatment-related negative side effects such as fatigue can be short-term (acute), persisting less than a month, or can be long-term (chronic), persisting from one month to years following treatment. Greater understanding of treatment-related negative side effects and possible interventions to alleviate these effects would lead to decreased hardships for patients during and following the cancer experience.

The authors became interested in using exercise as an intervention for cancer rehabilitation because of personal experiences. Susan Carter had watched many of her cancer patients struggling with the aftereffects of cancer therapies. In fact, one of the authors of this text, Carole Schneider, was diagnosed with cancer and became a patient of Dr. Carter's. She received extensive external and internal radiation, which left her struggling with many side effects, especially fatigue and muscle weakness. There appeared to be no knowledge about how to overcome the negative effects of the radiation. We began exploring the literature to find answers. While investigating possible interventions to alleviate Carole's symptoms, we found a close correspondence between the negative effects of cancer treatment toxicities and the positive effects that are observed with exercise. Investigating further, we found that there was minimal information on postcancer treatment interventions and that most of the existing information concerned the treatment of psychological effects. We began asking ourselves, "What interventions could alleviate the physiological side effects of cancer treatments?"

Since we were interested in the effects of exercise, we decided to begin the Rocky Mountain Cancer Rehabilitation Institute (RMCRI) to help cancer survivors regain their former quality of life through exercise interventions. Carole was the first cancer survivor in the Institute, and after two years of extreme fatigue and muscle weakness she began to regain her health and feel "well" again. Since that time we have worked with hundreds of cancer survivors and have found significant improvements, both physiological and psychological, in our patients. Consistently, we find that patients during and following cancer treatment improve their functional capacity, muscular strength and endurance, range of motion, pulmonary function, balance, agility, and blood profiles with concomitant decreases in fatigue and depression following an individualized prescriptive exercise intervention.

Since we could not find information in the professional literature, we decided to write this book to share our experiences at the Institute and to help you work with cancer survivors to improve their quality of life. We present scientific evidence from the assessments, exercise prescriptions, and exercise interventions that we have incorporated into our program; and we also give anecdotal information to help motivate and illuminate the application of these components. We want to emphasize that the individual cases should not be considered scientific evidence of the benefits of exercise for cancer patients. The scientific evidence comes instead from the research that we cite.

This is good news in the fight against cancer. New screening and diagnostic technology along with enhanced educational strategies has led to a significant reduction in cancer mortality. Millions have survived cancer; however, 72% to 95% of cancer survivors are experiencing cancer treatment-related problems. Rarely do you encounter a person who has not been touched by cancer. Cancer diagnosis can be devastating to the entire family as well as to the individual. It is paramount that we begin helping the millions of people who have survived cancer but are struggling to regain a life worth living.

Exercise and Cancer Recovery is a text for professionals working in cancer exercise rehabilitation in clinical and applied settings. The text can be used in undergraduate courses for training in cancer rehabilitation and in workshops training cancer exercise specialists. This text offers the reader a general overview of the etiology and associated effects of cancer. It includes background content to support specific guidelines for cancer exercise intervention and to provide information on rehabilitating the cancer patient during and following treatment. The text offers health care professionals (e.g., nurses, physical therapists, exercise physiologists, fitness specialists) the information they need in order to use exercise interventions effectively as a complementary therapy during and following treatment for cancer.

The text includes an overview of cancer pathology; the toxicities associated with the various cancer treatments; the information currently known on exercise as a therapy in cancer rehabilitation; and information on the assessments, exercise prescriptions, and exercise interventions (with examples) used at RMCRI that have led to significant improvement in the quality of life of cancer survivors. We include general suggestions on how to establish and manage a cancer rehabilitation facility.

Chapter organization is as follows:

• **Chapter 1** presents a brief overview of the causes of cancer pathologies and the latest statistical information on the widespread influence of cancer in the world. It includes a discussion of the process by which cancer cells develop and grow in human tissue and summarizes cancer types.

• **Chapter 2** contains information on treatments used to combat cancers and on the physiological effects these treatments have on human systems. The chapter relates these physiological effects to cancer treatment-related symptoms and outlines some of the effects of exercise.

• **Chapter 3** introduces the reader to the science of exercise physiology, including the principles of exercise and the physiological alterations that result from exercise intervention.

• **Chapter 4** offers information on the benefits of exercise for cancer survivors and introduces general guidelines on exercise intervention programs for cancer patients.

• **Chapter 5** presents assessment procedures and protocols used to help design exercise prescriptions for cancer patients. It also outlines modifications to the various assessments according to the health status of the patient.

• **Chapter 6** provides information on the specific components of the exercise prescription for cancer patients. Using one patient as a primary example, it provides a sample exercise prescription and sample workouts designed on the basis of the assessments discussed in chapter 5.

• **Chapter 7** gives information on exercise interventions for cancer survivors. It presents examples of the program developed for the patient described in chapter 6.

• **Chapter 8** offers information on management and patient care issues related to cancer rehabilitation, as well as information on the facilities and equipment that are needed to offer quality rehabilitation for cancer patients.

• **Chapter 9** briefly summarizes the text information on rehabilitation for cancer survivors.

The text incorporates features that will enhance students' understanding of the material. Included are periodic summaries of information, specific examples of the principles and programs, review questions, and a glossary of the terms that are italicized in the text. Another feature that readers should find valuable is a compilation of forms (appendix A) developed at RMCRI that readers may photocopy and use.

Acknowledgments

The authors would like to acknowledge and thank Mike Bahrke and Elaine Mustain for their dedication and efforts during the development of this book. The entire staff at Human Kinetics has been exceptional. We would also like to acknowledge the many contributions of our graduate students at the University of Northern Colorado, especially Ann Bentz, Barb Francis, Scott Drum, and Michelle Roozeboom.

1

Overview of Cancer Pathology

Despite advances in awareness, early diagnosis, and treatment, cancer is still among the diseases people fear most. In order to work effectively with individuals who have cancer, it is vital for health professionals to understand the disease. This chapter provides an overview of cancer pathology. We begin with a definition of cancer and then present statistics on cancer incidence. Subsequent sections deal with how cancer develops, types of cancers, and the grading and staging of tumors.

WHAT IS CANCER?

Various types of normal cells have various rates of growth and division. For example, once nerve cells mature, they grow no further, and cell division ceases. On the other hand, epithelial cells such as those in the skin or intestine grow and divide rapidly, continuously replacing other epithelial cells that have been lost. In some cases, however, the mechanisms that normally control the rapid growth and division of cells fail to perform that function effectively (DeVita, Hellman, and Rosenberg, 1997). Uncontrolled cell growth results in an abnormal mass of cells called a *tumor* (neoplasm) or in individual cells that are al-

tered to such a degree that they can no longer function normally (DeVita et al., 1997; Silverthorn, 1998; Snyder, 1986).

Tumors can be either *benign* or *malignant*. Benign tumors are slow growing, but can become quite large. They are well organized and well differentiated, and they do not have the destructive potential of malignant tumors. Benign tumors can damage adjacent areas and experience some areas of *necrosis* (tissue death), but they rarely cause death. Cancer cells are the cells of any malignant *neoplasm* (new growth). These cells are highly unorganized compared to normal cells. They have diverse abnormalities such as loss of cell *differentiation*, anomalous mitotic characteristics, increased invasiveness throughout the body *(metastasis)*, and decreased drug sensitivity. They are programmed for proliferation, grow on one another, invade normal tissue, and have the intrinsic ability to kill the host tissue (figure 1.1). Malignant tumors have poor definition and destroy the architecture of the invaded organ. Additionally, malignant tumors have large areas of necrosis due to lack of blood supply and lack of *apoptosis* (digestion by *phagocytes* of cell fragments from destroyed cells, a process that may be important in limiting the growth of tumors) (Snyder, 1986). "Cancer," then, is a general term for abnormal,

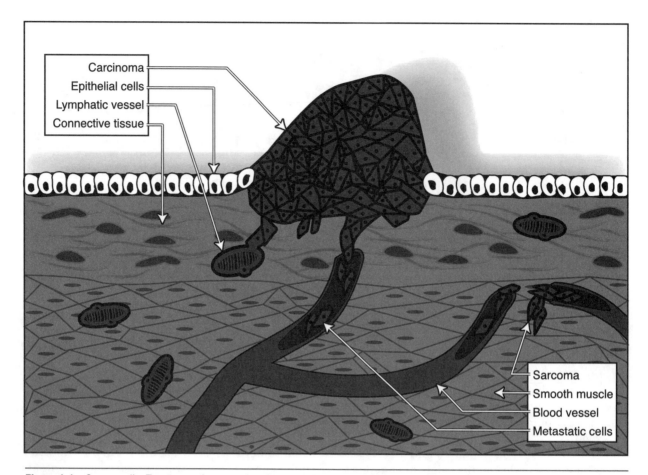

Figure 1.1 Cancer cells. Two types of cancers (carcinoma and sarcoma) are shown. Cells from the carcinoma are metastasizing (spreading) throughout the body.

uncontrolled cell growth that will lead to tissue failure unless the abnormal cells are removed or destroyed.

Metastasis is defined as the production of secondary malignant tumors originating from the primary tumor but located in anatomically distant tissues. All cancer has the potential to spread. Metastasis occurs when tumor cells detach from the primary tumor and enter the circulation or *lymphatics.* The cells then adhere to the subendothelial basement membrane of the distant tissue, which allows entrance into the new tissue. From this point on the cells proliferate and reproduce in the new tissue. Lung metastases are the most common secondary tumors in certain types of cancers, since the lungs are the first organs to filter malignant cells. But secondary tumors can also develop in other organs such as the liver and bones (table 1.1) (DeVita, Hellman, and Rosenberg, 2001; Dollinger, Rosenbaum, and Cable, 1997).

INTERNATIONAL CANCER STATISTICS

The World Health Organization estimates that in 2002 new cancer cases throughout the world will exceed 10 million, with 62% of these cases resulting in death (IARC Press, 2001). This estimate includes 4.7 million new cases in more developed countries and 5.4 million new cases in the less developed countries—that is, the number of new cancer cases will be approximately 13% greater in the less developed countries. Deaths in 2001 in more developed countries is approximately 2.6 million, while deaths from cancer for less developed countries is approximately 3.6 million, or 28% greater. International statistics by gender indicate that males have a higher incidence of cancer in more developed (2.5 million) and less developed (2.8 million) countries compared with females (2.1

Table 1.1   Common Metastatic Sites for Several Cancers

Type of cancer	Metastatic sites*
Breast	Brain, bone, lung
Cervix	Bladder, rectum, lung, liver, brain
Colorectum	Liver
Liver	Lung, bone
Lung	Brain, bone
Ovary	Liver, intestines, stomach, bladder
Pancreas	Liver, lung
Prostate	Bone
Sarcomas of bone	Lung
Stomach	Liver
Testis	Lung

*Secondary metastases sites are common for many types of cancers.

million and 2.5 million, respectively). Additionally, males have correspondingly higher mortality rates. Internationally, the most prevalent cancers for males and females are lung cancer and breast cancer, respectively. International statistics on cancer incidence, survival, and mortality for each country can be obtained from organizations such as the World Health Organization and the International Agency for Research on Cancer (IARC) (IARC Press, 2001).

The good news is that for most of the top 10 cancers, overall incidence and mortality have been declining slightly since 1990. In the United States, the National Cancer Institute estimates that there are approximately 8.9 million Americans who have survived cancer. In England the news is also positive, with the number of people surviving more than five years improving an average of 4% every five years. Nearly half of women and a third of men diagnosed now with cancer will live for at least five years, with cancer survival rates improving each year. The decline in death rates can be attributed to early detection and advancements in cancer treatment. For example, the death rate among women with breast cancer has decreased probably because of wider use of mammography.

Cervical cancer mortality rate has declined because of regular screening with the Papanicolaou smear test (Pap test). Colorectal cancer mortality rate has declined due to advanced sigmoidoscopy screening (Ries et al., 2000; United Kingdom Department of Health, 2001).

HOW CANCER DEVELOPS

Our bodies consist of a vast number of cells, all of which contain pairs of chromosomes. Deoxyribonucleic acid molecules (genes) spiral throughout each pair of chromosomes, giving the cell the "blueprint for life." The genes are continuously sending messages to the chromosomes, telling the body how to grow and function. Usually the genes function normally and send correct messages; however, "mistakes" can occur during reproduction or as a result of external factors. The body has the ability to repair some of the damage; but if the "mistake" occurs during the process of cell division, mutations can develop in one or more of the cell's genes. The genetic change or damage results in an abnormal chromosome within the cancer cell. Cancer develops as the abnormal chromosome starts receiving the wrong messages from the gene(s). The cells begin to grow rapidly and form tumors. Tumors develop in various organs and tissues, thus giving rise to cancer classifications.

Normal Cell Growth and Division

Understanding the process of cancer development requires an understanding of the structure, replication, and division of a normal cell. Briefly, the two major parts of a cell are the *nucleus* and the *cytoplasm*. The nucleus is the control center, governing the chemical reactions within the cell and the reproduction of the cell. Located within the nucleus of the cell are large quantities of *deoxyribonucleic acid (DNA)* molecules, which are called genes. Genes not only control heredity from parents to children but also control the day-to-day reproduction and function of all cells (Guyton and Hall, 2001; Silverthorn, 1998).

Somatic (body) cells can incur damage, disease, or destruction, leading to the promotion of new cell development to replace them. An adult loses billions of cells daily from various parts of the body. Some cells reproduce continuously (bone marrow, skin cells) while other cells either rarely or never reproduce (neurons, striated muscle cells). The process of cell division is generally an ordered process that is described by the *cell cycle*. Cell cycles vary in length from 8 hr to years.

The cell cycle consists of two phases: *interphase* and cell division *(mitosis or M phase)*. Interphase is the time during which no cell division is occurring and the cell carries out normal cellular processes of growth and metabolism. During interphase, which is 90% of the cell cycle, the cell is preparing for cell division. Interphase involves three distinct events: *synthesis,* or the S phase, and two *gap phases,* G_1 and G_2 (figure 1.2). The sequence of phases is G_1, S, and G_2. During G_1, cells are metabolically very active and involved in rapid protein synthesis. The S phase entails the synthesis and replication of DNA in the nucleus (chromosome replication) in preparation for cell division. G_2 is the phase when the cell increases protein synthesis, further preparing for cell division. Some cells that never reproduce (nerve cells) remain in the G_1 phase. Once a cell is in the S phase, it is committed to cell division (Rhoades and Pflanzer, 1996; Silverthorn, 1998).

There are a number of hypotheses for cell growth regulation. One hypothesis involves the

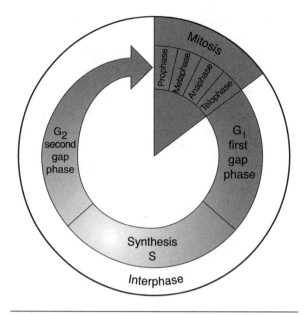

Figure 1.2   Cell cycle interphase. The longest cell cycle phase is interphase. Interphase comprises two gap phases and the synthesis phase. Mitosis occurs for a much shorter time period.

presence of growth factors. It appears that in some cases, growth factors within the blood or from adjacent tissues must be present before cells can grow. For example, epithelial cells in the pancreas will not grow without growth factor from the connective tissue of the gland. A second hypothesis suggests that the inhibition of cell growth may be due to lack of space. When a cell comes into contact with a solid object, the cell will stop growing. The last hypothesis involves cell growth inhibition due to a collection of the cell's own secretions—growth is controlled by the cell's own feedback system (DeVita et al., 1997; Rhoades and Pflanzer, 1996; Snyder, 1986; Tortora and Grabowski, 1993).

How Normal Cells Become Cancerous

Carcinogenesis is defined as the process by which a normal cell becomes cancerous. This process appears to occur in two stages (initiation and promotion) with a number of substages (figure 1.3). During the pre-initiation period, the genetic components of the chromosomes are protected from the effects of carcinogens. The assault to the *genome* of the cell occurs during the first stage *(ini-*

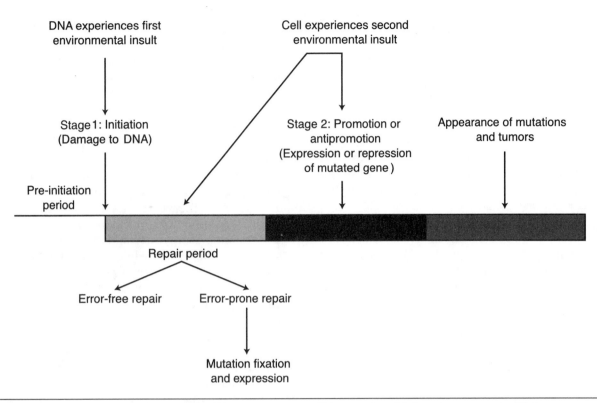

Figure 1.3 The process whereby a normal cell becomes cancerous. This general model shows how a normal cell becomes attacked by carcinogens. The cell tries to repair itself; if the repair is unsuccessful and the cell divides, the mutated gene is expressed resulting in uncontrolled cell division. DNA = deoxyribonucleic acid.

tiation), when a *carcinogen* attacks a normal cell. The resultant mutant cell is capable of sustaining uncontrolled cell division. Either the cell's DNA is altered by direct damage to the DNA molecules, or the cell's DNA repair system becomes inhibited and repair cannot occur. Before a cell becomes cancerous, as many as 10 distinct mutations may have to accumulate in the cell. Once the cell divides, the DNA mutation is expressed. The second stage *(promotion)* begins when the gene is expressed within the cell. The cell then begins uncontrolled division and tumor promotion. The genetic information can be expressed or repressed within the cell. If the mutated gene is repressed, normal function can occur but expression is always possible. Internal or external factors influence whether a mutated gene will be expressed or repressed (DeVita et al., 1997, 2001; Snyder, 1986).

Recently, the "multiple hit theory" of how cancer develops has gained popularity. According to this theory, cancer arises as a result of at least two "hits" or insults to the genes in the cell. The hits accumulate until cancerous growth is initiated. The hits may come from chemical or foreign substances (carcinogens, e.g., tobacco smoke), or they may be promoters (e.g., alcohol) that increase growth of abnormal cells. The number of hits, the types of hits, the frequency of hits, and the intensity of the hits are critical to the development of cancer. Accordingly, cancer is thought of as an additive process with multiple hits accumulating over a period of many years (Knudson and Strong, 1972).

An exciting and relatively new area of research on the origins of cancer deals with the heritability of some cancers. Early-onset breast and ovarian cancer are examples of inheritable cancers. Family members who are closely related to a patient with these cancers have approximately two to three times more risk of developing these cancers than does the general population. Thus, the identification of particular mutated genes (e.g., BRCA1 and/or BRCA2, associated with breast cancer) could have a significant impact on the prevention of certain cancers. Identification could result in early detection and preventative treatments. Only about 5% to 10% of cancers are considered hereditary (Dollinger et al., 1997; Marcus et al., 1996).

TYPES OF CANCER

Tumors are named according to the tissue or organ of origin and the degree of differentiation. The broad classes of cancer types include *carcinomas, melanomas, sarcomas, leukemia,* and *lymphomas.*

• **Carcinomas** are solid tumors that originate in the epithelial cells (lining of all tissues). Most epithelial carcinomas develop in organs that secrete substances. For example, lung tissue secretes mucus, breast tissue secretes milk, and the pancreas secretes digestive juices. Carcinomas can occur in the lungs, ovaries, breasts, kidneys, esophagus, stomach, uterus, and intestinal tract. Many types of carcinomas can occur within each tissue or organ. Carcinomas of the lung, for example, include adenocarcinoma, squamous cell carcinoma, and large cell carcinoma. The location, severity of prognosis, and histological properties delineate the type of lung cancer. Approximately 85% to 90% of all cancers are carcinomas.

• **Melanomas** are cancerous growths of melanocytes (malignant pigment-producing cells) that may be found throughout the body. Melanomas most commonly occur on the skin but can also arise in the back of the eye and in the mouth, vagina, and anus. Melanomas can spread outwardly on the surface of the skin or vertically, going deep into layers of tissue. Melanomas can be found initially internally with no sign of external skin involvement. Although melanomas usually contain black pigment, they can also be amelanotic, which means they contain no pigment.

• **Sarcomas** are solid tumors that originate in connective tissue, bone, muscle, cartilage, or fat. There are approximately 56 different types of sarcomas. For example, the four common types of bone sarcomas are osteosarcoma, Ewing's sarcoma, chondrosarcoma, and fibrosarcoma. The rarest form of cancer, sarcomas account for less than 2% of all cancers.

• **Leukemia** is a cancer of blood-forming organs that results from abnormal white blood cell (leukocyte) production in the bone marrow. Leukemia is classified as acute or chronic and according to the types of cells involved. Acute leukemia involves immature cells that interfere with the production of normal mature cells seen in the bone marrow and blood. Chronic leukemia involves mature white blood cells (rarely red blood cells) that appear normal but do not function normally and are unable to fight infection. Leukemia con-

stitutes about 2% of all cancers. Acute and chronic lymphoblastic leukemias, for example, involve abnormal, immature white blood cells that crowd the bone marrow and prevent red blood cell and platelet production and decrease production of normal white blood cells. Leukemic cells invade vital organs such as the liver, lungs, heart, and brain, in some cases resulting in organ dysfunction.

• **Lymphomas** are malignant cancers of the lymphocytes (type of white blood cells). The result is enlargement of the lymph glands and other organs in which lymphocytes normally develop. These cancers originate in the lymphatic system, which is located throughout the body and contains small vessels that filter foreign substances, especially bacteria, from the lymph fluid. Lymphoma accounts for approximately 5% of cancers and comprises about 20 different types of tumors. All persons with human immunodeficiency virus (HIV) infection are at increased risk to develop lymphoma, and this risk increases with time. There are two distinct groups of lymphomas:

- **Non-Hodgkin's lymphomas** are usually widespread within the lymph system and involve clonal (descended from a single cell) malignant expansion of B or T cells.

- **Hodgkin's lymphomas** progress in an orderly fashion from one lymph node group to the next group in a predictable spread. In this type of lymphoma, abnormal, multiform lymphocytes and other cells infiltrate the lymph nodes. Hodgkin's lymphoma and the lymphomas occurring in persons with HIV are high-grade B-cell lymphomas (Dollinger et al., 1997; Snyder, 1986).

GRADING AND STAGING OF TUMORS

The appropriate treatment for a particular cancer depends on the evaluation of the malignancy. The two types of evaluation are *grading* and *staging:*

Grading

Grading classifies tumor cells according to their microscopic appearance, which indicates the degree of undifferentiation *(anaplasia)* present in the cells. The less differentiated the cells, the more malignant the cancer. Thus, a grade I (low-grade) tumor has cells that are well differentiated and that resemble normal cells, and is usually slow growing and less aggressive, while grade IV tumors are composed of immature, poorly differentiated cells that make it difficult to determine the tissue of origin. Grade IV tumors are fast growing and aggressive. Grade II and III tumors have moderately differentiated and poorly differentiated cells, respectively. It appears that the cells that are the most able to reproduce and the least able to differentiate "outgrow" more normal cells. Thus the cancer becomes progressively more deadly (Snyder, 1986).

TNM Staging

The anatomic extent of the cancer must be determined before therapy can be prescribed for the cancer patient. The TNM staging system (table 1.2) for clinical and pathological evaluation is commonly used to determine the extent and progression of cancer. The higher the stage the further the cancer has progressed. The factors used to stage the cancer tumor are

- the local tumor size (T),
- the spread of the cancer to regional lymph nodes (N), and
- the presence or absence of distant metastasis (M).

Within the TNM system, four stages have been identified:

- Stage I denotes a mass limited to the organ of origin, no lymph node involvement, and no metastasis.
- In stage II the original tumor has spread into immediately surrounding tissue, and there is some lymph node involvement.

- Stage III tumors show an extensive primary lesion with fixation to deeper structures, and lymph nodes exhibit malignant invasion.
- In stage IV, distant metastases beyond the local site of the primary tumor are evident.

"Staging of Small Cell Carcinoma of the Lungs" on page 8 presents the TNM system for that type of cancer.

Other Staging Systems

There are a number of other staging systems that differ in the number of stages (three or four), clinical criteria versus pathologic criteria, tumor marker value inclusion, and use of size and sites of the primary tumor and metastatic disease. Staging can differ for different types of cancer. The TNM system meets the criteria developed by the American Joint Committee on Cancer and the Union Internationale Contre le Cancer. Thus the TNM system is the recommended classification system (Fleming et al., 1998).

Lymphoma Staging

The TNM system cannot be used for staging malignant lymphomas since there is no way to differentiate the T, N, and M in this type of cancer. Instead, lymphomas are grouped into four stages dependent on age, Ann Arbor stage (developed in 1970; includes subcategories for stages I through IV), the number of extranodal sites involved, performance status, and serum lactate dehydrogenase concentration. As an example, "Staging of Large Cell Non-Hodgkin's Lymphoma" on the next page shows the most widely accepted staging system for one type of lymphoma.

Table 1.2 Basic Staging of the TNM System

Tumor stage	Tumor size	Lymph nodes?	Metastasis
I	<2 cm	None	None
II	2-5 cm	No, or yes on same side	None
III	>5 cm	Yes on same side	None
IV	Does not matter	Does not matter	Yes

Staging of Small Cell Carcinoma of the Lungs

Primary Tumor (T)

T0   No evidence of primary tumor

T1   Tumor 3 cm or less in greatest dimension, surrounded by lung or visceral pleura, without bronchoscopic evidence of invasion more proximal than the lobar bronchus

T2   Tumor with any of the following features of size or extent:

      More than 3 cm in greatest dimension

      Involves main bronchus, 2 cm or more distal to the carina

      Invades the visceral pleura

      Associated with atelectasis or obstructive pneumonitis that extends to the hilar region but does not involve the entire lung

T3   Tumor of any size that directly invades any of the following: chest wall (including superior sulcus tumors), diaphragm, mediastinal pleura, parietal pericardium; or tumor in the main bronchus less than 2 cm distal to the carina, but without involvement of the carina; or associated atelectasis or obstructive pneumonitis of the entire lung

T4   Tumor of any size that invades any of the following: mediastinum, heart, great vessels, trachea, esophagus, vertebral body, carina; or separate tumor nodules in the same lobe; or tumor with a malignant pleural effusion

Regional Lymph Nodes (N)

N0   No regional lymph node metastasis

N1   Metastasis to ipsilateral peribronchial and/or ipsilateral hilar lymph nodes and intrapulmonary nodes, including involvement by direct extension of the primary tumor

N2   Metastasis to ipsilateral mediastinal and/or subcarinal lymph node(s)

N3   Metastasis to contralateral mediastinal, contralateral hilar, ipsilateral or contralateral scalene, or supraclavicular lymph node(s)

Distant Metastasis (M)

M0   No distant metastasis

M1   Distant metastasis present (Fleming et al., 1998)

Staging of Large Cell Non-Hodgkin's Lymphoma

Stage I   Involvement of a single lymph node region or localized involvement of a single extralymphatic organ or site

Stage II   Involvement of two or more lymph node regions on the same side of the diaphragm, or localized involvement of a single associated extralymphatic organ or site and its regional nodes with or without other lymph node regions on the same side of the diaphragm

Stage III   Involvement of lymph node regions on both sides of the diaphragm, which may also be accompanied by localized involvement of an extralymphatic organ or site, by involvement of the spleen

Stage IV   Disseminated (multifocal) involvement of one or more extralymphatic organs with or without associated lymph node involvement, or isolated extralymphatic organ involvement with distant (nonregional) nodal involvement (Fleming et al., 1998)

Leukemias

Leukemias are not staged but are placed into classifications for determination of treatment methodology. The classification is based on morphologic examination for lineage (descent), confirmation of lineage by cytochemical (reaction of cells to chemical agents) stains, quantification of the number of blasts, and estimation of the degree of differentiation of the cells (DeVita et al., 1997).

FURTHER READINGS

DeVita, V.T., S. Hellman, and S.A. Rosenberg. 2001. *Cancer: Principles and practice of oncology* (6th ed.). Philadelphia: Lippincott Williams & Wilkins. Comment: This text presents detailed information on the principles of cancer including molecular biology, genetics, cell invasion, and metastases. The information is very current with a high level of detail but is technically very difficult.

Dollinger, M., E.H. Rosenbaum, and G. Cable. 1997. *Everyone's guide to cancer therapy* (3rd ed.). Kansas City, MO: Andrews McMeel. Comment: This book provides basic information on understanding cancer. It is written in simple language; some of the information is somewhat outdated.

Fleming, I.D., J.S. Cooper, D.E. Henson, R.P. Hutter, B.J. Kennedy, G.P. Murphy, B. O'Sullivan, L.H. Sobin, and J.W. Yarbro (eds.). 1998. *American Joint Committee on Cancer staging handbook* (5th ed.). Philadelphia: Lippincott-Raven. Comment: Detailed information on the staging and grading of cancers is the focus of this book. The information is current but is technically difficult.

Guyton, A.C., and J.E. Hall. 2001. *Textbook of medical physiology* (10th ed.). Philadelphia: Saunders. Comment: This text presents information on normal cell division and replication. Material is very scientific but easily understandable.

SUMMARY

Cancer is a term for abnormal, uncontrolled cell growth that results in tissue failure. Uncontrolled cell growth can lead to the development of tumors that are either benign or malignant. Metastasis, or the production of secondary malignant tumors, can develop in distant tissues if the primary tumor goes undetected and untreated. Statistical information details cancer as a health issue prevalent in all societies. Males have a higher incidence of and mortality from cancer compared to females. The most common male cancer is lung cancer, and the most common female cancer is breast cancer. The good news, however, is that the mortality rate for most cancer types is declining.

Normal cell growth and division involve a number of cell cycle phases. Each cell cycle is regulated by a number of factors such as growth factors, spatial availability, or regulation by the cell itself. When carcinogenesis occurs, the once-normal cell becomes cancerous. A carcinogen (i.e., environmental pollutant) attacks a normal cell, causing it to become mutated. The mutated cell is capable of uncontrolled cell division and tumor development. The tumor is named according to the tissue or organ where it originated. Basic classes of cancer types include carcinomas, melanomas, sarcomas, leukemia, and lymphomas. The most common tumors are carcinomas; sarcomas are the rarest. The treatment for the various types of cancers is dependent on the evaluation of the tumor. Tumors are graded and staged to determine their severity and progression. Grading involves determination of the microscopic appearance of the tumor; staging involves clinical and pathological evaluation. The most common and widely used staging method is called the TNM system, whereby the tumor size (T), the extent of the involvement of regional lymph nodes (N), and the presence or absence of distant spreading or metastasis (M) are determined. Various types of cancers are staged in different ways.

STUDY QUESTIONS

1. Provide a useful definition for the term cancer.

2. What is the difference between a benign and a malignant tumor?

3. Describe the normal process of somatic cell division (mitosis).

4. Explain the various hypotheses for cell growth regulation.

5. How do normal cells become cancerous?

6. Distinguish between the major classifications or types of cancer.

7. Explain the TNM system for clinical evaluation of cancerous tumors.

REFERENCES

DeVita, V.T., S. Hellman, and S.A. Rosenberg. 1997. *Cancer: Principles and practice of oncology* (5th ed.). Philadelphia: Lippincott-Raven.

DeVita, V.T., S. Hellman, and S.A. Rosenberg. 2001. *Cancer: Principles and practice of oncology* (6th ed.). Philadelphia: Lippincott Williams & Wilkins.

Dollinger, M., E.H. Rosenbaum, and G. Cable. 1997. *Everyone's guide to cancer therapy* (3rd ed.). Kansas City, MO: Andrews McMeel.

Fleming, I.D., J.S. Cooper, D.E. Henson, R.P. Hutter, B.J. Kennedy, G.P. Murphy, B. O'Sullivan, L.H. Sobin, and J.W. Yarbro (eds.). 1998. *American Joint Committee on Cancer staging handbook* (5th ed.). Philadelphia: Lippincott-Raven.

Guyton, A.C., and J.E. Hall. 2001. *Textbook of medical physiology* (10th ed.). Philadelphia: Saunders.

International Agency for Research on Cancer. 2001. *GLOBOCAN 2000: Cancer incidence, mortality and prevalence worldwide.* IARC CancerBase No. 5. Lyon, France: IARC Press.

Knudson, A.G. Jr., and L.C. Strong. 1972. Mutation and cancer: A model for Wilms' tumor of the kidney. *J Natl Cancer Inst* 48:313.

Marcus, J.N., P. Watson, D.L. Page, S.A. Narod, G.M. Lenoir, P. Tonin, L. Linder-Stephenson, G. Salerno, T.A. Conway, and H.T. Lynch. 1996. Hereditary breast cancer: Pathobiology, prognosis, and BRCA1 and BRCA2 gene linkage. *Cancer* 77(4):697–709.

Rhoades, R., and R. Pflanzer. 1996. *Human physiology* (3rd ed.). Philadelphia: Saunders College.

Ries, L.A.G., P.A. Wingo, D.S. Miller, H.L. Howe, H.K. Weir, H.M. Rosenberg, S.W. Vernon, K. Cronin, and B.K. Edwards. 2000. Annual report to the nation on the status of cancer. *Cancer* 88(10):2398–2424.

Silverthorn, D.U. 1998. *Human physiology: An integrated approach.* Upper Saddle River, NJ: Prentice Hall.

Snyder, C.C. 1986. *Oncology nursing.* Boston: Little, Brown.

Tortora, G.J., and S.R. Grabowski. 1993. *Principles of anatomy and physiology* (7th ed.). New York: Harper Collins College.

United Kingdom Department of Health. 2001. *The NHS cancer plan.* London, England: Department of Health.

2

Cancer Treatment Toxicity on Physiological Systems and the Benefits of Exercise

Exercise specialists working with cancer patients need to understand the principles of various treatments, as well as the physiological toxicities that occur with these treatments, in order to safely and effectively assess patients and prescribe appropriate exercise interventions. The exercise intervention is based on the type of cancer treatment the patient is receiving, in addition to the stage and progression of the disease.

It is often the case that cancer treatments successfully kill cancer cells, but the patient pays for this outcome with unpleasant side effects. Treatment toxicities have the potential to affect almost every physiological system. This chapter introduces you to the treatments for cancer, from those that are the most widely used (surgery, radiation, and chemotherapy) to those that are still experimental. We then present information on the major side effects of cancer treatment, including the general effects of the therapies and the ways in

which these therapies damage specific systems in the body.

CANCER TREATMENTS

Treating cancer can be extremely difficult because not all cancer cells behave in the same way. As suggested in chapter 1, cancer is often not a single, specific disease. By the time a cancerous tumor is detectable, it may contain a diverse population of cells. For example, some cells divide and others do not; some cells metastasize whereas others do not; and some cells are highly responsive to drug intervention while other cells are not.

Cancer treatments include numerous therapeutic modalities. The treatments we cover in this section comprise common methodologies and experimental treatments: surgery, radiotherapy, chemotherapy, hormonal therapy, immunotherapy (*antibody* therapy), bone marrow transplantation, gene therapy, and vaccine therapy. Gene therapy and vaccine therapy are in the experimental stages of development. The type and technique of therapy used, alone or in combination with another treatment, are decided on the basis of factors such as response rate, drug sensitivity, and side effects.

Surgery

Surgery in relation to cancer has three main roles: prevention, diagnosis, and treatment.

Surgery for Prevention and Diagnosis

Although within the context of cancer the non-medical public usually associates surgery with treatment, surgery is often vital in prevention and diagnosis as well.

• **Prevention.** Certain conditions, known as precancerous conditions, are associated with a high risk that the person will develop a particular cancer. When these conditions exist, it may be necessary to remove an organ or a growth that could become malignant. For example, polyps in vascular organs such as the uterus or rectum may become malignant. Another example is skin anomalies that could become malignant. In extreme cases, for example when an individual has a serious familial history of breast or ovarian cancer, removal of the organs likely to be affected may be a reasonable prophylactic measure. The benefits of prophylactic surgery must be carefully balanced with the risks of such surgery and the permanent changes that will result from the surgery (DeVita, Hellman, and Rosenberg, 2001).

• **Diagnosis.** The role of surgery (invasive procedures) in the diagnosis of cancer mainly involves obtaining tissue for *histological examination.* Many techniques may be used to obtain tissue for analysis, such as aspiration biopsy, core biopsy, punch biopsy, and incisional and excisional biopsy. In *aspiration biopsy,* fragments are removed from the suspect tissue by collection through a needle. In core biopsy, a specially designed needle is inserted into a tissue, and a cylindrical sample of the tissue is removed. The biopsy core can usually be used for diagnosis of tumor type. With punch biopsy, a shallow fragment of tissue is removed. In *incisional biopsy,* a small wedge of tissue is removed from a larger mass for diagnosis. In *excisional biopsy,* the entire suspected tumor with minimal surrounding normal tissue is excised. The technique of choice should be the technique that gives the most adequate sample for pathological analysis. The pathological analysis helps in staging the cancer for appropriate treatment.

Surgery As Treatment

Surgery can be a safe, effective method of curing cancer in patients with solid tumors that are confined to one tissue of origin. The role of surgery in the treatment of cancer falls into one of six areas:

1. **Definitive surgical treatment for removal of the primary cancer with selection of appropriate local therapy.** The challenge is to select the appropriate local therapy after surgery. Local therapy differs with the type of cancer and the site of involvement. In many cases, surgical removal of both the cancer and some normal surrounding tissue is sufficient local therapy. In other cases, surgery may need to be integrated with adjuvant therapies (chemotherapy, radiation, etc.).

2. **Surgery to reduce the bulk of the disease.** The purpose of reducing the bulk of the disease is to give the adjuvant therapies (chemotherapy, radiotherapy, etc.) a better chance for success. Removing a carcinoma along with the primary lymph nodes in the area is one example.

3. **Surgical resection of metastatic disease.** In addition to treating the primary cancer, if metastasis is confined to one organ, then surgery can be used to remove the metastatic disease. An example is surgery for brain cancer in a case in which the brain is the only site of known metastatic disease.

4. **Surgery due to emergency situations caused by the cancer.** Examples include surgery to correct hemorrhage or perforation.

5. **Surgery to decrease disease-related symptoms to provide comfort for the patient and not for the purpose of curing the patient.** At this point the cancer is incurable and the aim of palliative surgery is to reduce pain or functional abnormalities. Examples are the removal of intestinal obstructions and the removal of tumors that are causing pain.

6. **Surgery for reconstruction and rehabilitation.** The aim of surgery in this area is usually to improve function or cosmetic appearance, for example in reconstructive surgery

after a mastectomy or muscle transplantation to restore muscle function damaged by radiation (DeVita et al., 2001; Dollinger, Rosenbaum, and Cable, 1997).

Surgical treatment, then, aims to remove the cancerous tumor cells and some surrounding healthy tissue while causing minimal injury to the normal functioning of the affected area. Surgically decreasing the population of tumor cells increases the chances that other combinations of treatments (radiotherapy, chemotherapy, etc.) will be effective. Surgery can be very successful for slowly proliferating cancers with long cell cycles and for cancers with cells that are highly cohesive and slow to metastasize.

Radiation Therapy

Radiation therapy (radiotherapy) uses high-energy X rays, electron beams, or radioactive isotopes for the purpose of damaging or destroying malignant cancer cells. The objective of radiotherapy is to destroy tumor cells with minimal injury to normal tissue and organs. Radiotherapy is used to target small areas and is not typically used for metastatic cancers. There are two types of radiotherapy: external and internal. *External radiotherapy* aims beams of radiation from a source outside the body at the targeted tissue within the body; *internal radiotherapy* places radioactive isotopes near the tumor within the body.

Once it has been ascertained that radiotherapy is the appropriate treatment, the precise location of the target cancer cells, the best beam distribution, and the transit route that will most effectively minimize normal tissue damage must be determined. The direct target of radiation is the cancer cells' DNA (deoxyribonucleic acid), whose molecular structure can be directly altered. Radiation can also affect the cell membrane. Additionally, photons may interact with the water within the cell and produce free radicals (atoms with impaired electrons and thus are not bound to another chemical compound; they can be harmful). During or following radiation, cancer or normal cells can show one or more of the following effects:

- Necrosis (cell death)
- Apoptosis (cell deletion)
- Accelerated cellular *senescence* (aging)
- *Terminal differentiation* (a cell becomes sterile, or unable to divide, thus losing its poten-

tial to change physical and functional properties) (DeVita, Hellman, and Rosenberg, 1997; DeVita et al., 2001).

As mentioned in chapter 1, the cell cycle is divided into G_1, S, G_2, and M (see figure 1.2, p. 4). Within both normal and tumor cells, the M phase of the cycle is the most sensitive to radiation, followed by G_2. The greatest resistance to radiation occurs in the late S phase (Terasima and Tolmach, 1963). The response to radiation in normal and cancer cells depends on cell proliferation. Cell damage in the radiated area is exacerbated if the tissue is experiencing cell proliferation, thereby increasing the number of cells exposed to the radiation. The number of cells in the tumor or in the surrounding normal tissue increases because of

- enhanced reproduction of cells in the tumor or in normal tissues or
- recruitment of new cells from nonradiated areas.

Enhanced reproduction of cells occurs when cells within the tumor reproduce rapidly or when normal tissues, in response to damage, produce growth factors that stimulate cell reproduction to aid in the repair process. Recruitment of new cells from nonradiated areas occurs to help in damage prevention and repair. Enhanced reproduction and recruitment of cells from other areas bring more cells into the localized radiation area for potential necrosis, thus contributing to the negative side effects experienced by cancer patients. Additionally, immediately after radiation there appears to be a release of a necrosis factor and of growth factors that may add to local damage or contribute to distant *cytotoxic* effects. For example, platelet growth factor is released from the vascular endothelium, stimulating proliferation of smooth muscle in the small *arterioles* and thus contributing to long-term vascular problems. The good news is that tissues that function without constant cell reproduction (i.e., muscle and neural tissue) are resistant to radiation. However, the vascular and connective tissues that support muscle and nerves do reproduce often and thus are susceptible to radiation. Other tissues that reproduce frequently (skin, gastrointestinal mucosa, bone marrow, reproductive tissue, and exocrine glands) are also extremely sensitive to radiation. If tissues have slow cell proliferation (i.e., lungs), the effects of radiation are seen much later. Further on in the chapter, in "Cancer Treatment-Related Side Effects"

(p. 17), we discuss in more detail the effects of radiation on normal cells.

Radiation and surgery complement each other because their effectiveness is in different locations within the cancer tumor area. Radiation is most effective at the periphery of the tumor where there are few cells. Surgery is most effective in the center of the tumor where there are large numbers of cancerous cells. Radiation can be administered before or after surgery. The advantages of radiation before surgery include the possibility of minimizing the tumor size and of rendering the cells sterile. The major disadvantage of administering radiation before surgery is the potential that radiation could alter the tumor enough to affect the pathology report. Another disadvantage is that alteration of the tumor prevents accurate staging. Finally, if *fibrosis (scarring)* is extensive, surgery can be technically very difficult (DeVita et al., 1997, 2001; Dollinger et al., 1997).

Chemotherapy

Chemotherapy is used in a number of ways: as the primary therapy for advanced cancer, as an adjuvant therapy with other localized treatments, and as a primary therapy to control localized cancer.

Chemotherapy is the administration of antitumor drugs that destroy malignant tumor cells (Dollinger et al., 1997; Tenenbaum, 1994). As with radiation therapy, the goal is not only to destroy tumor cells but also to minimize the destruction of normal cells and limit the disruption of normal cell function. But because normal cells and cancer cells are quite similar, it is very difficult to find drugs that will destroy the tumor without affecting normal cells. Remember, cancer cells have characteristics that differentiate them from normal tissue cells, but a cancer cell was potentially a normal cell at one time. Therefore, there are similarities between normal and cancer cells. For instance, both types of cells have a nucleus and cytoplasm; however, the structure of these differs. Recently, drug treatments for leukemia have given rise to some hope for the development of drugs that will "attack" only cancer cells and not cause normal cell injury. The way in which many chemotherapy agents work, however, does give normal cells one advantage. The *antineoplastic* effects of chemotherapy (effects that interfere with the growth of malignant cells) are most pronounced in cells during the proliferation phases of the cell cycle; and because cancer cells are proliferating

more rapidly than normal cells, chemotherapy drugs destroy higher percentages of cancer cells than of normal cells.

Chemotherapy agents used in the treatment of a variety of types of cancer are grouped into categories that include alkylating agents, antimetabolites, antitumor antibiotics, and alkaloids.

- *Alkylating agents* attack all cells in a tumor, whether they are reproducing or not. These agents bind with the DNA in the cells to prevent reproduction.
- *Antimetabolites* attack the cells during cell division. These chemotherapy drugs imitate normal cell nutrients so that the cell consumes the drug but eventually starves to death.
- *Antitumor antibiotics* insert into strands of DNA, either breaking the chromosomes or inhibiting the synthesis of RNA (ribonucleic acid), which plays an important role in synthesis within cells.
- *Alkaloids* prevent cell duplication by interrupting the formation of chromosomes.

Although chemotherapy treatment may consist of a single drug or a combination of antitumor drugs from various categories, using a combination of drugs seems to be more effective. Combination chemotherapy can provide a number of benefits:

- Maximum cell necrosis from each drug (each of which can be better tolerated by normal tissue because the dosages are lower)
- A broader range of coverage against various types of resistant cells
- A slowing of the development of cancer cells that are resistant to any particular therapy (Chabner and Longo, 2001)

Chemotherapy drugs can be administered orally in pill, capsule, or liquid form. *Intravenous* delivery can be via a single venipuncture per treatment or through temporary catheters placed in the skin. Implanted infusion ports are catheters that are implanted totally under the skin. A needle is used to inject the chemotherapy drug(s) into the chamber of the port. Ambulatory pumps are small portable pumps that deliver the chemotherapy drug during normal daily activities (figure 2.1); the administration and delivery rate of the chemical agents are controlled by a battery. Another method is central nervous system delivery by injection directly into the spinal fluid. Some leukemia patients need chemotherapy administration through spinal taps. The *intraperitoneal* method delivers the drug directly into the abdominal cavity through a catheter or a port into a compartment located on the chest or abdominal wall that releases the drugs in the abdominal cavity. All is placed into the abdomen during surgery. Chemotherapy drugs can also be administered through the artery that supplies the infected organ via a pump that can overcome arterial pressure. Infusion pumps are yet another method of administration. These pumps, implanted surgically, can be used to deliver chemotherapy drugs repeatedly over extended periods of time.

Figure 2.1 Example of a chemotherapy pump. The route of administration is determined by the type of tumor and the drug being used.

Chemotherapy drug administration may occur in the hospital, in the physician's office, or at home, depending on the administration method. The length of treatment varies with the treatment program, but six months is a typical time period for a chemotherapy regimen. A cycle of therapy may consist of administration of one drug every day for two weeks with other drugs administered on particular days during the two-week period. Following a period during which normal cells have time to recover, the cycle is typically repeated every few weeks throughout the six-month program (Dollinger et al., 1997).

Radiation and chemotherapy can be used together. Chemotherapy can increase the response of cells to radiation, allowing better local control of the cancer. Additionally, the two therapies can be combined so that the radiation targets the local tumor while the chemotherapy is aimed at eliminating micrometastases (small quantities of metastatic cells).

As a cancer exercise specialist, it is imperative for you to be aware of the patient's regimen of radiation, chemotherapy, or both. The type of agent, the schedule, and the means of delivery are all significant in varying degrees and at varying times in relation to decisions that you will make about the patient's exercise program.

Hormonal Therapy

Hormonal agents are used in treating cancers such as breast, prostate, or endometrial carcinomas. The high hormone levels in these tissues (e.g., of estrogen in the breast) affect the growth of cells, whether normal or cancerous.

The aim of hormonal therapy is to change the response of the hormone or the response of its receptor. For example, the antiestrogen tamoxifen prevents gene activation of estrogen *receptors,* blocks the reproduction of cells during the G_1 cell cycle phase, and inhibits DNA enzymatic activity. Tamoxifen is usually taken in 20-mg doses for up to five years. Tamoxifen has been effective as an adjuvant treatment for breast cancer, has been shown to be effective in treating metastatic breast cancer, and reduces the risk of breast cancer recurrence approximately 40%. Other hormonal agents have been successful in hormonally responsive tissues. For example, prostate cancer is treated with antiandrogens, which interfere with androgen receptor activation and inhibit DNA synthesis in the prostate (Chabner and Longo, 2001; McTiernan, Gralow, and Talbott, 2000).

Immunotherapy

The immune system identifies and responds to foreign substances introduced into the body. These unwanted agents include pathogenic organisms, bacteria, viruses, and cancer cells that can alter normal cells. To facilitate an effective defense, the immune system features complex processes that identify the "enemy" and coordinate the defense against it. The immune defense system involves humoral immune responses and cellular immune responses. *Humoral immune responses* defend against foreign invaders outside the cellular structures (e.g., within the lymph system). Relying on *antibodies* derived from the *B lymphocytes* for antigen recognition and recruitment of phagocytes, humoral immune responses attack extracellular pathogens (e.g., tumor cells in the lymph system). Within the cell itself, whether within tumor cells or normal host cells containing tumor cells, *cellular immune responses* try to defend the cell from intracellular pathogens by recognizing and destroying tumor cells. Cellular immune responses rely on the *T-cell* receptor for recognition of cell-surface *antigens* and for triggering T-cell activities that destroy the tumor cells directly or indirectly through *macrophages*.

Immunotherapy uses agents called *biological response modifiers*. Biological response modifiers usually have the ability to alter the immunological host tissue response when the tumor invades, or they can alter the tumor itself. These agents (a) employ the body's own defense mechanisms to combat *tumorogenesis* or (b) facilitate radiation and chemotherapy through regeneration of normal bone marrow or by stimulating the production of antibodies and/or lymphocytes to destroy tumor cells. Examples of biological agents include *immunoaugmenting agents* such as interferons, which are involved in the disruption of viral replication, limiting the advancement of viral establishment in neighboring cells. *Immunoregulating agents* such as interleukins, which activate B cells and T cells or stimulate growth of T cells, enhance the inflammatory response by summoning macrophages and *natural killer cells*. *Immunorestorative agents* inhibit the blocking of suppressor T cells and tumor necrosis factors produced by macrophages and lymphocytes; can stimulate immune cells; directly damage tumor cells; affect the blood vessels around tumors; and make tumor cells more identifiable to the immune system (Chabner and Longo, 2001).

Bone Marrow Transplantation

In a bone marrow transplant, diseased bone marrow is replaced by healthy bone marrow. Bone marrow transplantation has been used to treat cancers such as leukemias, Hodgkin's disease, large cell non-Hodgkin's disease, other cancer lymphomas, and some solid tumor types of cancers such as testicular cancer and breast cancer if the patient cannot be cured by a simpler therapy. Transplants can be allogenic, syngeneic, or autolo-

gous. In an *allogenic* transplant, bone marrow is transferred from one individual to another. These individuals can be siblings or can be persons who are unrelated to each other. *Syngeneic* transplants use identical twins, who have the same bone marrow genetic makeup, as donor and recipient. Unless the patient has an identical twin, finding matched donors is very difficult. Therefore, *autologous* transplants are used more often than either allogenic or syngeneic transplants. The patient's bone marrow is removed and stored until treatment is complete. Following high-dose chemotherapy, radiation, or both, the patient's own bone marrow is reimplanted. Although this eliminates the risk of rejection of the infused bone marrow, there is a risk of reinfusing cancer cells that were stored along with the bone marrow. The transplantation for a leukemia patient can be either allogenic or autologous; however, with the use of autologous transplants, there must be a source of uncontaminated bone marrow cells. This can be accomplished by trying to eliminate cancer cells in the person's bone marrow collected during a period of remission (Dollinger et al., 1997).

Experimental Therapies

Numerous experimental therapies have been introduced to combat the many types of cancer disease. Gene therapy began in 1989 with the first gene transfer trial. Since then, clinical trials for many gene therapy approaches have been approved by the National Institutes of Health. The information being obtained from clinical trials will hopefully produce a new generation of cancer drugs that will manipulate genetic factors to control cancer. Another experimental approach to the prevention and treatment of cancer is vaccine therapy. Since 1975, many oncogenic (giving rise to tumors) infectious agents have been identified, but clinical trials so far have produced limited data. Mostly the use of vaccines has centered on the prevention of infectious diseases linked to the development of cancer.

Gene Therapy

As discussed in chapter 1, cancer can develop as a result of genetic damage or loss. Gene therapy inserts a functioning gene into a mutated cell to correct the abnormality or to perform a new function that will enable the cell to survive. This can involve a number of strategies, such as the following:

- Genetic modification of the immune response, for example active immunization and genetic modification of *immune effector cells,* such as those that increase tumor recognition

- Modification of cancer tumors with genes that have direct antitumor effects, such as tumor suppressor genes

- Introduction of genes into hematopoietic stem cells to decrease toxicity from chemotherapy

- *Antiangiogenic* gene therapy involves genetic modification to permit formation of new blood vessels

There are numerous methods for gene transfer into cells. Because viruses can deliver genes to cells with high efficiency, many gene transfer techniques use viruses to introduce genetic material. However, nonviral transfer methods have fewer safety issues, and it is easier to produce clinical-grade material for nonviral cell transfers (DeVita et al., 2001).

Vaccine Therapy

The two types of cancer vaccines are preventive cancer vaccines and therapeutic or treatment vaccines. Preventive cancer vaccines aim to attenuate the cancerous activity of the infectious agent (bacterium, virus) by reducing or preventing the infection of the target tissue (stomach, cervix, liver). The aim of therapeutic cancer vaccines is to activate the immune system to destroy already existing cancer cells. Clinical trials are addressing the potential benefits and risks of vaccine therapy (DeVita et al., 2001).

CANCER TREATMENT-RELATED SIDE EFFECTS

As noted earlier, in the process of destroying cancer cells, radiation therapy and chemotherapy also cause physiological alterations to normal tissue and body functions. Severe side effects include toxicities in many body systems and organs. These toxicities are dependent on choice of therapy, dose, and the patient's tolerance. In this section we outline the general effects of these therapies and then consider damage to specific systems in the body. In discussing the systems we include information about the effect of the toxicity on a patient's quality of life.

Radiation therapy kills cells most effectively when they are undergoing mitosis (cell division); normal tissues that function without constant cell reproduction (e.g., muscle, neural, and lung tissue) are more resistant to damage from radiation than are cancer cells, which are rapidly and constantly dividing. However, both the normal vascular and connective tissues that support muscle and nerves do reproduce frequently (though not as frequently as cancer cells do) and thus are more susceptible to radiation than less active normal tissue. Other tissues that reproduce frequently (skin, gastrointestinal mucosa, bone marrow, reproductive tissue, and exocrine glands) are also extremely sensitive to radiation. General damage that can be caused by radiation therapy includes

- damage to the cell membrane, causing changes in the membrane transport system; and

- anemia as a result of radiation-induced hemorrhage.

Radiation has *acute* (immediate) and *chronic* (late) effects on human systems. Acutely, reproducing and repairing cells are affected. Chronically, necrosis, fibrosis, *fistula* formation, ulcerations, and damage to specific organs occur. These chronic effects appear to depend on the total dose of radiation and on the length and number of radiation treatments.

Chemotherapy, like radiation therapy, is most effective against cells that are reproducing. It follows that normal tissues that have continuous turnover of cells—such as bone marrow, which is continuously producing red blood cells, white blood cells, and platelets—are more likely to experience toxic effects with chemotherapy treatment than are those tissues that proliferate slowly.

Immune System Toxicity

Radiation therapy and chemotherapy in high doses have a profound effect on the immune system. The tissues (B lymphocytes and T cells) so prominent in the immune system are particularly vulnerable to damage during and following cancer therapy. Immune system cells that reproduce following cancer therapy, especially radiation, are much more susceptible to cell death. The *hematopoietic tissues* (tissues that produce blood cells, i.e., bone marrow) are vulnerable as well. There can be effects on bone marrow stem cells, causing myelosuppression, which results in tissues that can no longer perform their blood cell-creating function effectively. The likely effects are

- *leukopenia* (low levels of white blood cells in general);

- *lymphocytopenia* (low levels of white blood cells produced in the lymph system);

- *granulocytopenia* (deficiency of granulated white blood cells: neutrophils—the first line of defense against infection; basophils—cells that bring anticoagulant substances to inflamed tissues; and eosinophils—cells that detoxify inflammation-inducing substances); and

- *thrombocytopenia* (decreased numbers of circulating mature platelets).

The extent of cell death and *myelosuppression* (inhibition of bone marrow function) caused by damage to bone marrow stem cells depends on the type, dose, and localization of the cancer therapy (DeVita et al., 1997; Kohn and Melvold, 1976).

Cancer therapy-induced myelosuppression can leave the cancer patient highly susceptible to infections. Cancer patients often have respiratory problems, colds, cough, and flu-like symptoms. Medications that they receive to combat these infections leave them feeling even more tired and exhausted. One of our patients at the Rocky Mountain Cancer Rehabilitation Institute (RMCRI) said, "It is bad enough having to fight cancer but also having a cold all the time makes life even more difficult." Patients must limit the environments that they are exposed to—for instance, they must stay away from their children's school when the school population has excessive cases of colds and flu. We do not allow RMCRI cancer exercise specialists who have a cold or any flu-like symptoms to work with cancer patients. Patients with thrombocytopenia must be careful not to get any scratches or cuts since bleeding is a problem. Additionally, these patients bruise easily if bumped. Most cancer patients at RMCRI have continuous monitoring of their blood parameters through their physician. Those with low platelets often have excessive bruising at the point of venipuncture and experience excessive swelling and pain in the affected arm.

Cardiovascular Toxicity

Radiation has significant acute and chronic side effects on the cardiovascular system. Although not all the tissues that experience most of the acute effects of radiation (the basal cells of the epidermis, the mucosal epithelia, and the hematopoietic cells of the bone marrow) are directly involved in the cardiovascular system, the injury that they sustain leads secondarily to the release of histamines that cause inflammatory responses in the vasculature. This inflammation leads to vascular dilation, increased capillary permeability, and interstitial edema. In addition, the vasculature may experience a long-term effect due to radiation-stimulated release of *platelet-derived growth factor* and *fibroblast growth factor* with resulting cell proliferation within the small arterioles. This leaves the vascular tissue vulnerable to damage or destruction due to lack of oxygen and nutrients. Long-term effects on the vascular system worsen with progressively reduced capillarization in the irradiated tissue, resulting in decreased blood *perfusion* in the tissue. *Anemia* (low levels of red blood cells resulting in low concentrations of hemoglobin and decreased hematocrit levels) occurs rather quickly because of damage to hematopoietic tissue. For example, investigations have shown that acute *hemolytic* anemia can occur within six days of radiation exposure (cited in Haylock and Hart, 1979). But outright hematopoietic failure is delayed because bone marrow failure occurs gradually and because red blood cells are long-lived. These secondary responses may be as damaging for the cancer patient as the primary responses (DeVita et al., 2001; Hellman and Botnick, 1977; Reinhold and Buisman, 1973).

The central structures of the cardiovascular system are also seriously affected by radiation treatment. With thoracic (chest) radiotherapy, *pericarditis* (inflammation of the membranous sac enclosing the heart) is the most common side effect. Some patients recover from pericarditis in two to five months following treatment, while others require surgery to prevent restrictive inflammation surrounding the heart. Radiation-induced pericarditis is fatal in about 6% of patients. Other reported abnormalities include cardiac conduction problems developing 6 to 23 years following radiation and reduced ventricular function 5 to 15 years after radiation (DeVita et al., 2001; Fischer, Knobf, and Durivage, 1993).

Chemotherapy agents (e.g., doxorubicin, 5-fluorouracil) can also directly damage the heart. There appears to be a gradual increase in heart damage with increased doses. The cardiomyopathy that occurs with chemotherapy appears in about four weeks following chemotherapy and has a 50%

higher fatality rate than does radiation-induced cardiomyopathy. Signs indicating chemotherapy-induced cardiac damage include sinus tachycardia, premature atrial and ventricular beats, and supraventricular arrhythmias (Becker, Erckenbrecht, Haussinger, and Frieling, 1999; Brockstein, Smiley, Al-Sadir, and Williams, 2000; Fischer et al., 1993; Gianni et al., 2001; Keefe, Roistacher, and Pierri, 1993; Nousiainen, Vanninen, Rantala, Jantunen, and Hartikainen, 1999; Weinstein, Mihm, and Bauer, 2000).

Cancer patients' quality of life is severely compromised if they are experiencing any cardiovascular or blood-related side effects. Anemia is common in cancer patients, resulting in extreme fatigue and tiredness. Lower red blood cell counts, of course, mean reduced oxygen-carrying capacity. Fatigue results as the body is forced to work harder even during normal daily activities. Irregular heartbeats can be very disturbing to cancer patients. At RMCRI we had a patient who was having arrhythmias but was cleared to remain in our program after doing a stress test in which the cardiologist could not detect any unusual heart problems. The patient stated, "I want to stop the exercise program, my irregular heart rate is very disturbing." It is difficult to determine whether the cardiovascular problems that our patients experience are a consequence of their cancer treatment or of cardiovascular disease independent of their cancer treatment.

Pulmonary Toxicity

Pulmonary toxicity may be acute or chronic, developing within days, months, or years following radiotherapy and chemotherapy. Long-term treatment may cause intra-alveolar pulmonary fibrosis (formation of scar tissue) and abnormal development of pulmonary tissue—especially the *endothelial cells* and *epithelial cells.* Patients may experience coughing, *dyspnea,* and a low-grade fever as a result of pulmonary toxicity. Additionally, patients have a diminished diffusion capacity and decreased pulmonary compliance from eight months to 10 years after therapy. Pulmonary toxicity also presents itself as fatigue, low exercise tolerance, restlessness, and *tachypnea* (Brockstein, et al., 2000; Chabner and Longo, 2001; Fischer et al., 1993; Wilson, 1978).

Cancer patients who experience pulmonary toxicities have excessive shortness of breath, frequent colds, low functional capacity, and fatigue.

One of our cancer patients at RMCRI was told by his doctors that the exercise program at RMCRI was "the only reason he was doing as well as he was." Another patient who was experiencing lung and breathing difficulties improved her functional capacity 10-fold. She said, "This program has added so much to my quality of life, I could hardly breathe and could not do even daily activities without sitting down frequently." Her husband stated that if his wife had not been in our program he felt "he would have lost her a long time ago."

Gastrointestinal System Toxicity

Cancer therapies can cause acute changes in the intestinal mucosa, probably due to cell reproductive cycle damage. Chronic intestinal changes include thickening of some bowel segments; *stenosis* (narrowing) of the bowel; ulceration; intestinal fibrosis; vascular edema and wall thickening; distortion of the arteries; and increased intestinal *motility* leading to diarrhea. Additionally, levels of *disaccharidases* and *aminopeptidases* (the digestive enzymes involved in the catabolism of carbohydrates and proteins) decrease, and the absorption capacity of the intestine is reduced. Both of these conditions can lead to protein deficiency and fat, carbohydrate, vitamin, and electrolyte absorption impairments. Abdominal radiation and chemotherapy commonly produce vomiting, nausea, and loss of appetite. Although there is no means of preventing all of these symptoms, highly effective anti-nausea medications allow the patient to continue with cancer therapy. Malnutrition is an obvious danger, as are reduced energy production and dehydration. Extreme fatigue, as with other toxicities, is often the result. The loss of appetite is often compounded by stress, anxiety, and depression. Fortunately, loss of appetite is usually short-term (Chabner and Longo, 2001; DeVita et al., 2001; Fischer et al., 1993).

Cancer patients with gastrointestinal toxicities are often extremely thin and weak because of malnutrition. Patients also have stomach pain; can't keep food down; lose their sense of taste; experience abdominal pain, constipation, or diarrhea; have blood in their stools; and experience fatigue. Some patients have colostomies and don't want to go out in public. One of our patients at RMCRI was so weak that he had to stop working as a mail carrier. When he entered our program he could hardly walk, but six months later he was running at least 30 min per training session. Needless to say, his functional aerobic capacity had increased, as had his muscular strength. His levels of fatigue and weakness decreased significantly.

Musculoskeletal Alterations

Radiation and chemotherapy disturb muscle integrity. Radiotherapy may alter the sarcolemma, sarcoplasmic reticulum, and mitochondrial membranes, leading to disturbances in muscle force generation. Evidence also suggests that the myofibrils and myofilaments incur damage and disorganization. Acute changes in the connective tissue, muscular tissue, and *serous membrane* include edema, hemorrhage, inflammation, and vascular spasms (DeVita et al., 2001; Fischer et al., 1993). Cancer patients experience extreme muscular weakness, fatigue, muscle imbalances, and decreased range of motion. Our patients at RMCRI have shown significant improvements in these symptoms from their exercise programs.

The muscle weakness that results from musculoskeletal changes is often very profound. When our patients first come to us, they commonly express deep unhappiness at their degree of debilitation. One of our patients stated, "I am so frustrated by my weakness; all I want to do is be able to carry my children like other mothers or put clothes in my washer like I once could." Other patients often say things like "I am so stressed, I can't open a jar, walk upstairs, or lift anything like I used to be able to."

Hepatic Toxicity

Some chemotherapy agents, such as methotrexate, are hepatotoxic. It appears that these agents cause liver injury leading to hepatocyte necrosis, *steatosis* (fatty degeneration), and *cholestasis* (arrest of bile excretions). Chronic alterations to the liver include cytotoxic lesions, hepatic fibrosis, cirrhosis (disease of the liver), and chronic cholestasis (loss of bile excretion). Liver disease is usually temporary and can be controlled with careful therapy planning (Chabner and Longo, 2001; DeVita et al., 1997; Fischer et al., 1993).

Acute hepatic (liver) radiation toxicity usually presents within two to six weeks after completion of radiation, although it may not manifest itself for six months or longer. Liver enzymes are also elevated in the blood serum. Patients who incur hepatic toxicity may experience rapid weight gain, increase in abdominal girth, fatigue, anorexia, and pain. One of our patients undergoing treatment with a combination of drugs including methotrexate said, "I was so fatigued when I entered your program, I am so excited, I have energy now."

Neuroendocrine Toxicity

High doses of radiotherapy can affect thyroid tissue, leading to cell necrosis and atrophy. Radiation-induced DNA damage and a continuous stimulation by thyroid-stimulating hormone cause major thyroid malfunction. Radiotherapy can also affect the hypothalamus and pituitary, causing abnormalities in the release of growth hormone.

Chemotherapy-induced neurotoxicity includes a wide range of central nervous system symptoms, from cognitive dysfunction to urinary incontinence and blurred vision. The incidence of neurologic side effects is low to moderate, and depends on factors such as dose and age of the patient. Symptoms may persist up to several years following treatment (Chabner and Longo, 2001; Fischer et al., 1993; Snyder, 1986).

Cancer patients often experience confusion, memory loss, seizures, numbness, decreased sensation, foot drop, muscle weakness, constipation, urinary incontinence, blurred vision, hearing loss, vestibular problems (loss of balance), and peripheral neuropathy. Susan, an RMCRI patient, stated, "I have chemo brain, I can't think, I can't remember anything, and I often feel like I am going to fall." Balance problems are severe in many of our patients. We do a lot of balance exercises. Peripheral neuropathy appears in the patients' extremities, with loss of neural control resulting in foot drop, reduced grip strength, and the like. We have found that exercise leads to improvements in our patients' grip strength and balance.

Nephrotoxicity

Some chemotherapy agents, such as cisplatin, mitomycin, and methotrexate, may cause hyperuricemia (abnormal amount of uric acid in the blood), resulting in gout; kidney and bladder abnormalities; hemolytic anemia; edema; and decreased magnesium, calcium, potassium (hypokalemia), and sodium. Fluid intake and medications such as mannitol and amifostine have been effective in controlling *nephrotoxicity* in cancer patients (DeVita et al., 1997; Fischer et al., 1993; Snyder, 1986).

Patients experience swelling and pain from gout, painful edema or swelling in the upper and lower limbs, and dehydration. At RMCRI we have had patients with severe lymphedema in the arms and legs. We monitor this swelling carefully and have not found that it is exacerbated with exercise. Our patients are also very dehydrated (have very dry skin), so water consumption is encouraged at all times.

Dermatological Toxicity

Because chemotherapy drugs fail to differentiate between actively reproducing cancer cells and normal cells and because hair cells actively reproduce, these drugs destroy hair cells. The result is hair loss. Chemotherapy agents can also affect the skin so that it forms lesions, especially in a radiated area.

Patients experience skin infections, head dermatitis from wigs and the excessive heat caused by wigs, and embarrassment and loss of the sense of self. Anne stated, "I am embarrassed to go out in public with no hair." She rarely left her house, and when she did she always wore a hat that totally covered her head.

A CLOSER LOOK AT SOME COMMON SYMPTOMS

The effects of the toxicities just described profoundly influence the quality of life of cancer patients. This section more fully addresses several of these more common side effects—often the result of toxicities in multiple systems. These side effects include fatigue, lymphedema, pain, body image problems, sleep disturbances, and various forms of psychological distress (most notably depression and anxiety).

Fatigue

Cancer patients experience a type of fatigue that is far more disruptive to their quality of life than in patients with almost any other disease. Fatigue has been described as overwhelming, whole-body tiredness, not related to activity or exertion, that makes a person feel drained and produces a strong desire to lie down and sleep. Most attempts to define fatigue scientifically involve behavioral/performance, physiological/biochemical, and sensory/subjective elements.

Prevalence of Fatigue

According to the National Cancer Institute (2002), 72% to 95% of patients receiving cancer therapy experience a debilitating fatigue resulting in diminished work capacity. Because of fatigue, patients discontinue treatment; doses of some treatments must be limited; and quality of life is impaired. Investigations (Irvine, Vincent, Graydon, Bubela, and Thompson, 1994; King, Nail, Kreamer, Strohl, and Johnson, 1985; Meyerowitz, Watkins, and Sparks, 1983) have shown that the majority of cancer patients experience fatigue regardless of the type or length of treatment. The Fatigue Coalition (1998b) found that 45% of patients experienced fatigue after one week of chemotherapy, while an additional 33% experienced fatigue two weeks after chemotherapy.

The majority of studies (Irvine et al., 1994; King et al., 1985; Meyerowitz et al., 1983) show that the prevalence of fatigue increases over the course of cancer treatment regardless of its type. The fatigue pattern varies over the course of, and with the type of, therapy:

• **Chemotherapy.** Pickard-Holley (1991) found that in most patients, fatigue occurred 3 to 4 days following chemotherapy, became more severe 10 days following chemotherapy, and then declined until the next cycle of chemotherapy. The physical fatigue and other flu-like symptoms such as fever, chills, headache, and malaise exacerbate feelings of overall fatigue (Haeuber, 1989).

• **Surgery.** Fatigue has been reported following surgery. Cancer patients who have surgery and chemotherapy or radiation may find that their fatigue is accumulative. Christensen, Hjorts, Mortensen, Riis-Hansen, and Kehlet (1992) found that postoperative fatigue was physiological in nature and did not have psychological determinants.

• **Radiation therapy.** Greenberg, Sawicka, Eisenthal, and Ross (1992) found that fatigue reached a maximum in the fourth week (average of 17 treatments) in women (mean age of 46 years) with stage I or II node-negative breast cancer who were receiving radiation treatment after lumpectomy for local disease control. Again, the researchers found that the patients' fatigue was biological in nature and was independent of depressive symptoms.

Perceptions of Effects of Fatigue

The Fatigue Coalition (1998a) conducted a survey among 419 cancer patients and 205 physicians, as well as among caregivers of cancer patients. Here are some of the results from the patients and the physicians:

• **Patients.** Seventy-eight percent experienced fatigue significant enough to affect their daily living habits. Sixty-one percent stated that fatigue affected their lives more than pain did. The patients related that the aspects of their lives that were affected included their physical well-being (60%), work capacity (61%), emotional status (51%), enjoyment of life (57%), and intimacy with their significant loved one (44%). In this study, the patients reported that they had to limit their social activity (57%), had problems finishing tasks (49%), and could not walk far without becoming fatigued (48%). Twelve percent wanted to die because they felt so fatigued, and 16% felt that treating their fatigue was as important as treating the cancer.

• **Physicians.** In contrast to the patients surveyed, only 26% of the physicians felt that fatigue affected their patients more than pain did. Likewise, the physicians did not agree with patients as to the cause of fatigue: physicians (55%) believed that cancer causes fatigue, while patients and caregivers felt that fatigue was caused by the treatments (54% and 61%, respectively). This may be one of the main reasons that 75% of the patients felt they must just accept their fatigue and did not tell their physician about this debilitating side effect. Only 35% of physicians said that they offer treatments for fatigue to their patients.

The discrepancies between the patients and the physicians in this survey make it clear that you, as a health care worker in oncology, must take the time to listen to your patients and must do all you can to promote the team approach to treatment (with the team including the patient).

Physiological Mechanisms That Cause Fatigue

A number of known mechanisms have the potential to contribute to the fatigue of cancer patients:

• As noted previously, the diarrhea, vomiting, and loss of appetite resulting from gastrointestinal toxicity lead to fatigue as a consequence of the loss of body nutrients, fluid, and electrolytes (Gordeuk and Brittenham, 1992; Groopman and Itri, 1999; Resbeut et al., 1997; Seifert, Nesser, and Thompson, 1994).

• Neurotoxicity can slow motor function and cause fatigue.

• Cardiotoxicity affects the function of the cardiovascular system. More stress on the heart leads to fatigue. The stress on the heart is a result of left ventricular dysfunction, increased time to peak filling of the left ventricle, lower left and right ejection fractions, abnormal left ventricular contractility, reduced cardiac output and stroke volume, and lower oxygen and nutrient delivery to the organs during normal daily activities.

• Pulmonary toxicity can result in decreased total lung capacity, vital capacity, inspiratory capacity, and diffusion capacity, which lead to lower functional or work capacity.

• Hepatotoxicity and nephrotoxicity affect the liver and kidney functions, leading to interference with metabolic function and a decrement in energy production (Fischer et al., 1993).

• Finally, the destruction of red blood cells has an adverse effect on oxygen delivery to exercising tissues, and this effect in turn is a major contributor to exhaustion.

Although a great deal remains to be learned about physiological mechanisms of fatigue, it is clear that long-term decrements in muscular strength, cardiac function, pulmonary function, red blood cell count, and motor function—resulting from both the disease and its treatment—lead to a loss of work capacity and to fatigue.

Psychological Factors in Fatigue

Psychological factors and fatigue are clearly linked in cancer patients. But the relationship appears to vary from patient to patient, and the mechanisms involved are not well understood. The literature reveals diverse observations on this issue:

• The National Cancer Institute (2002) reports that approximately 40% to 60% of fatigue in cancer patients does not have disease or physical causes but relates to the moods, beliefs, attitudes, anxiety, and depression consequent to the emotional stress of cancer diagnosis.

- Hayes (1991) states that during cancer, depression and fatigue may occur together as a result of the same biological alterations.

- Cognitive problems such as difficulty thinking, forgetfulness, inability to concentrate, and decreased attention are commonly associated with fatigue in cancer patients (Rhodes, Watson, and Hanson, 1988).

- Piper (1990) questioned women who were receiving treatment for breast cancer. The patients indicated that changes in their psychological state (stress, worry, depression, and anxiety) contributed the most to their feelings of fatigue.

- The mental demands made on cancer patients throughout diagnosis and treatment are well documented (Loveys and Klaich, 1991). These demands could produce attentional fatigue.

Even though there is no overarching explanation of how psychological factors affect fatigue in cancer patients and vice versa, it is clear that psychological factors can strongly influence the outcome of cancer treatment and that the disease can produce or exacerbate psychological problems. The psychological benefits of exercise include reductions in anxiety, tension, and depression; and these reductions contribute to the well-being of the active person. Moderate aerobic exercise reduces anxiety, muscle tension, and blood pressure for 2 to 5 hr after the exercise bout. Chronic exercise is also associated with low levels of anxiety, reduced muscle tension, and reduced depression (Bouchard, Shephard, Stephens, Sutton, and McPherson, 1990). The data thus far from RMCRI have shown significant reductions in depression with improved quality of life in our cancer patients.

Lymphedema

Lymphedema is swelling caused by a buildup of lymph fluid in soft tissues. It develops because of a blockage within the lymph system. Cancer patients experience lymphedema as a result of scarring after the surgical removal of lymph nodes or after radiation therapy. Large areas of lymph nodes (armpit, pelvic region, groin) are obstructed, and swelling in the arms and legs results. The swelling leads to excessive discomfort and pain. The functional ability of the swollen limb is also hindered. Exercise has a place in both the prevention and treatment of lymphedema. For example, if a patient has had surgery for breast cancer, regaining full range of motion and strength in the arm can help reduce the risk of lymphedema. Exercise is also used to manage lymphedema in breast cancer patients (McTiernan et al., 2000). At RMCRI we monitor the circumferences of the various limbs each week, and we have not found exercise to have any adverse effects in our cancer patients.

Pain

Cancers of the bone and the abdominal organs usually result in severe pain in the advanced stages. Patients can experience two types of pain. Somatic pain is described as dull and throbbing, while neuropathic pain is sharp and shooting. Chemotherapy drugs and radiation can cause pain in numerous ways. For example, chemotherapy medications can cause peripheral neuropathy, which manifests as a burning in the hands and feet. Radiation may induce pain by irritating skin or scarring nerves. Pain can lead to depression, loss of appetite, anger, loss of sleep, and an inability to cope. Cancer-related pain can be reduced through a regular program of physical activity. The pain reduction may result from better functional capacity and strength or from reduced emotional factors (McTiernan et al., 2000). Colleen, one of our patients at RMCRI, had severe pain and fatigue before beginning her program. She stated, "[Y]ou have given me my life back." Like many of our patients, she found that her exercise intervention greatly alleviated debilitating pain.

Body Image Problems

Body image changes may be difficult for patients. The patient must cope with physical losses ranging from loss of hair to loss of one or both breasts, the colon, jaw, or almost any other body part. The patient may experience extreme weight gains or losses from various chemotherapy agents or from the progress of the disease itself. Lymphedema may produce swelling sufficient to cause some patients to feel that they look grotesque. Actual or perceived body image changes usually precipitate grieving, anger, or depression. One of our patients at RMCRI stated, "I want to go swimming and yet I am concerned about my deformity [breast surgery], I feel people will be disgusted or repelled by my appearance." We will examine later how exercise can help with self-image, from improving a patient's appearance to helping improve the function of many damaged body segments.

Sleep Disturbances

Cancer patients may have sleep disturbances for physical or psychological reasons. The causative factors may include pain, shortness of breath, fear, and depression. Certain types of chemotherapy interventions and radiation can affect sleeping patterns. Determining the causative factor is important so that treatment can be administered to reduce the frequency and severity of sleep disturbances. A psychologist or counselor in your facility can work with the patient to determine the cause for loss of sleep. Exercise has been shown to improve sleeping patterns in breast cancer patients (McTiernan et al., 2000). Our patient who was so thankful that our program had given her back her life reported having a much easier time sleeping following her exercise program, as do many of our other patients.

Depression and Anxiety

Depression and anxiety afflict 20% to 50% of cancer patients (Pirl and Roth, 1999; Schwenk, 1998). Depressive symptoms are linked to a lack of physical health, a sense of helplessness, pain, and alterations in body image. Depression can cause emotional instability, produce a negative psychological outlook, reduce functional status, and influence the patient's medical outcome (Schwenk, 1998; Watson, Haviland, Greer, Davidson, and Bliss, 1999).

The severity of depression and anxiety fluctuates depending on the type of cancer, the stage of cancer, the cancer treatment, and the coping strategies of the patient. If the cancer patient is uncertain of the outcome, anxiety is more severe. The depressed patient feels unable to cope and is certain that others cannot help. Depression may be manifested as decreased communication, insomnia, lethargy, and fatigue.

Cancer and cancer treatment can also affect cognitive abilities. Patients can experience difficulty thinking, forgetfulness, inability to concentrate, and decreased attention. They also have mood swings, anxiety, stress, and worry. Detailed psychological alterations are beyond the scope of this text; there are many comprehensive texts on psychological and social changes with cancer diagnosis. Regular exercise improves mood and alleviates anxiety, tension, anger, and depression. Studies on women with breast cancer have shown that those who are physically active have an improved quality of life and lower anxiety and depression (McTiernan et al., 2000). Our patients at RMCRI have shown improved physiological functioning with concomitant reductions in depression and fatigue. One of our patients who was very depressed about her cancer said, "I can't believe how exercising and working with the cancer specialists have made me feel so much more hopeful and happy."

FURTHER READINGS

Chabner, B.A., and D.L. Longo. 2001. *Cancer chemotherapy and biotherapy* (3rd ed.). Philadelphia: Lippincott Williams & Wilkins. Comment: This book provides technical information on chemotherapy and chemotherapy drugs, as well as extensive information on side effects.

DeVita, V.T., S. Hellman, and S.A. Rosenberg. 2001. *Cancer: Principles and practice of oncology* (6th ed.). Vols. 1 and 2. Philadelphia: Lippincott Williams & Wilkins. Comment: These books present highly technical information on many treatments and some of their side effects.

Dollinger, M., E.H. Rosenbaum, and G. Cable. 1997. *Everyone's guide to cancer therapy* (3rd ed.). Kansas City, MO: Andrews McMeel. Comment: This text offers basic information on treatments and side effects. It is written in simple language; some of the information is somewhat outdated.

McTiernan, A., J. Gralow, and L. Talbott. 2000. *Breast fitness*. New York: St. Martin's Press. Comment: This book provides basic information on exercise for breast cancer patients and survivors. Information about cancer is limited to breast cancer.

SUMMARY

Clearly, the physiological and psychological changes that occur with the diagnosis of cancer have a profoundly negative effect on patients' quality of life. The dimensions of quality of life that are measured in cancer research include physical factors (function, symptoms), psychological (emotional) factors, participation in activities, social and interactional factors, financial factors, and global factors (combination of factors). Cancer fatigue and cancer treatment-related fatigue exacerbate the problems that patients experience in every one of these quality-of-life dimensions (Ferrell, Grant, Dean, Funk, and Ly, 1996; Mor, 1992). Patients at RMCRI have experienced these cancer- and cancer treatment-related symptoms with diminished quality of life. Within the six-month exercise intervention that RMCRI provides, all patients have reported significant improvements in their quality of life since they were diagnosed with cancer. If, then, we care about our patients—and obviously we do—we will realize how vital it is for us to incorporate exercise into patients' treatment regimens to help reduce the fatigue and other negative side effects that so markedly influence every aspect of their lives.

STUDY QUESTIONS

1. Explain the major medical treatment modalities to combat cancer.

2. Explain how each treatment modality works to destroy cancerous tumors.

3. Discuss specific physiological alterations that accompany the various treatment modalities.

4. How do treatment effects manifest themselves in fatigue symptoms?

5. Explain the other symptoms that can occur as a result of cancer treatments.

REFERENCES

Becker, K., J.F. Erckenbrecht, D. Haussinger, and T. Frieling. 1999. Cardiotoxicity of the antiproliferative compound fluorouracil. *Drugs* 57(4):475–484.

Bouchard, C., R.J. Shephard, T. Stephens, J.R. Sutton, and B.D. McPherson. 1990. *Exercise, fitness, and health*. Champaign, IL: Human Kinetics.

Brockstein, B.E., C. Smiley, J. Al-Sadir, and S.F. Williams. 2000. Cardiac and pulmonary toxicity in patients undergoing high-dose chemotherapy for lymphoma and breast cancer: Prognostic factors. *Bone Mar Trans* 25(8):885–894.

Chabner, B.A., and D.L. Longo. 2001. *Cancer chemotherapy and biotherapy* (3rd ed.). Philadelphia: Lippincott Williams & Wilkins.

Christensen, T., N. Hjorts, E. Mortensen, M. Riis-Hansen, and H. Kehlet. 1992. Fatigue and anxiety in surgical patients. *Br J Surg* 79:165–168.

DeVita, V.T., S. Hellman, and S.A. Rosenberg. 1997. *Cancer: Principles and practice of oncology* (5th ed.). Vols. 1 and 2. Philadelphia: Lippincott-Raven.

DeVita, V.T., S. Hellman, and S.A. Rosenberg. 2001. *Cancer: Principles and practice of oncology* (6th ed.). Vols. 1 and 2. Philadelphia: Lippincott Williams & Wilkins.

Dollinger, M., E.H. Rosenbaum, and G. Cable. 1997. *Everyone's guide to cancer therapy* (3rd ed.). Kansas City, MO: Andrews McMeel.

Fatigue Coalition. 1998a. Cancer and fatigue: A survey among physicians, patients and caregivers. Available at http://www.cancercare.org. Accessed May 5, 2000.

Fatigue Coalition. 1998b. Fatigue is most prevalent and longest-lasting cancer-related side effect. Available at http://www.cancercare.org. Accessed May 5, 2000.

Ferrell, B.R., M. Grant, G.E. Dean, B. Funk, and J. Ly. 1996. "Bone tired": The experience of fatigue and its impact on quality of life (pp. 1–14). Available at info@fatiguenet.com. Accessed November 1, 1998.

Fischer, D.S., M.T. Knobf, and H.J. Durivage. 1993. *The cancer chemotherapy handbook* (4th ed.). St. Louis: Mosby.

Gianni, L., P. Dombernowsky, G. Sledge, M. Martin, D. Amadori, S.G. Arbuck, P. Ravdin, M. Brown, M. Messina, D. Tuck, C. Weil, and B. Winograd. 2001. Cardiac function following combination therapy with paclitaxel and doxorubicin: An analysis of 657 women with advanced breast cancer. *Ann Oncol* 12(8):1067–1073.

Gordeuk, V.R., and G.M. Brittenham. 1992. Bleomycin-reactive iron in patients with acute non-lymphocytic leukemia. *FEBS Letters* 308(1):4–6.

Greenberg, D., J. Sawicka, S. Eisenthal, and D. Ross. 1992. Fatigue syndrome due to localized radiation. *J Pain Sympt Mgmnt* 7:38–45.

Groopman, J.E., and L.M. Itri. 1999. Chemotherapy-induced anemia in adults: Incidence and treatment. *J Natl Cancer Inst* 91(19):1616–1634.

Haeuber, D. 1989. Recent advances in the management of biotherapy-related side effects: Flu-like syndrome. *Oncol Nurs Forum* 16 (Suppl. 6):35-41.

Hayes, J.R. 1991. Depression and chronic fatigue in cancer patients. *Primary Care* 18:327–339.

Haylock, P.J., and L.K. Hart. 1979. Fatigue in patients receiving localized radiation. *Cancer Nurs* 2:461–467.

Hellman, S., and L.E. Botnick. 1977. Stem cell depletion: An explanation of the late effects of cytotoxins. *Int J Radiat Oncol Biol Phys* 2:181.

Irvine, D., L. Vincent, J.E. Graydon, N. Bubela, and L. Thompson. 1994. The prevalence and correlates of fatigue in patients receiving treatment with chemotherapy and radiotherapy: A comparison with the fatigue experienced by healthy individuals. *Cancer Nurs* 17(5):367–378.

Keefe, D.L., N. Roistacher, and M.K. Pierri. 1993. Clinical cardiotoxicity of 5-fluorouracil. *J Clin Pharmacol* 33(11):1060–1070.

King, K.G., L.M. Nail, K. Kreamer, R.A. Strohl, and J.E. Johnson. 1985. Patients' descriptions of the experience of receiving radiation therapy. *Oncol Nurs Forum* 12(4):55–61.

Kohn, H.I., and R.W. Melvold. 1976. Divergent x-ray-induced mutation rates in the mouse for hand "7 locus" groups of loci. *Nature* 259:209.

Loveys, B.J., and K. Klaich. 1991. Breast cancer: Demands of illness. *Oncol Nurs Forum* 18:75–80.

McTiernan, A., J. Gralow, and L. Talbott. 2000. *Breast fitness*. New York: St. Martin's Press.

Meyerowitz, B.E., I.K. Watkins, and F.C. Sparks. 1983. Quality of life for breast cancer patients receiving adjuvant chemotherapy. *Am J Nurs* 83(2):232–235.

Mor, V. 1992. QOL measurement scales for cancer patients: Differentiating effects of age from effects of illness. *Oncology* 6(2):146–152.

National Cancer Institute. 2002. Information from PDQ for patients. Available at http://www.cancer.gov/cancerinfo/pdq/supportivecare/fatigue/patient/. Accessed September 10, 2002.

Nousiainen, T., E. Vanninen, A. Rantala, E. Jantunen, and J. Hartikainen. 1999. QT dispersion and late potentials during doxorubicin therapy for non-Hodgkin's lymphoma. *J Intern Med* 245(4):359–364.

Pickard-Holley, S. 1991. Fatigue in cancer patients: A descriptive study. *Cancer Nurs* 14(1):13–19.

Piper, B.F. 1990. Fatigue. *Trans-cultural implications for nursing interventions.* Presentation at the Sixth International Conference on Cancer Nursing, Amsterdam.

Pirl, W.F., and A.J. Roth. 1999. Diagnosis and treatment of depression in cancer patients. *Oncology* 13(9):1293–1301.

Reinhold, R.S., and G.H. Buisman. 1973. Radiosensitivity of capillary endothelium. *Br J Radiol* 46:54.

Resbeut, M., P. Marteau, D. Cowen, P. Richaud, S. Bourdin, J.B. Dubois, P. Mere, and T.D. N'Guyen. 1997. A randomized double blind placebo controlled multicenter study of mesalazine for the prevention of acute radiation enteritis. *Radiother Oncol* 44(1):59–63.

Rhodes, V., P. Watson, and B. Hanson. 1988. Patients' descriptions of the influence of tiredness and weakness on self-care abilities. *Cancer Nurs* 11:186–194.

Schwenk, T.L. 1998. Cancer and depression. *Oncology* 25(2):505–513.

Seifert, C.F., M.E. Nesser, and D.F. Thompson. 1994. Dexrazoxane in prevention of doxorubicin-induced cardiotoxicity. *Ann Pharmacother* 28(9):1063–1072.

Snyder, C.C. 1986. *Oncology nursing.* Boston: Little, Brown.

Tenenbaum, L. 1994. *Cancer chemotherapy and biotherapy: A reference guide* (2nd ed.). Philadelphia: Saunders.

Terasima, R., and I.J. Tolmach. 1963. X-ray sensitivity and DNA synthesis in synchronous populations of HeLa cells. *Science* 140:490-492.

Watson, M., J.S. Haviland, S. Greer, J. Davidson, and J.M. Bliss. 1999. Influence of psychological response on survival in breast cancer: A populations-based cohort study. *Lancet* 354:1331–1336.

Weinstein, D.M., M.J. Mihm, and J.A. Bauer. 2000. Cardiac peroxynitrite formation and left ventricular dysfunction following doxorubicin treatment in mice. *Pharm Exper Ther* 294:396–401.

Wilson, J.K.V. 1978. Pulmonary toxicity of antineoplastic drugs. *Cancer Treat Rep* 62:2003.

3

Fundamentals of Exercise

Because of the effects of regular exercise in reducing the risks of cardiovascular disease, improving pulmonary function, reducing obesity, and minimizing the impacts of aging, our society has been sent a strong message—*be active and stay active.*

This message applies as much to cancer patients as to anyone else—perhaps even more. An aim of this text as a whole is to show how individualized exercise programs can help overcome or reduce many of the side effects of cancer treatments. To explain how exercise can influence cancer treatment effects, this chapter outlines the fundamental principles of exercise, providing basic concepts for those who have no background in exercise physiology (oncology nurses, physical therapists, etc.). Readers who are in the field of exercise physiology will find the material a basic review. A final section of the chapter summarizes the fundamental principles of exercise prescription.

The best way to ensure that exercise produces desired individual outcomes is to recognize the needs, limitations, and capacities of the individual and then customize programs to fit the unique combination of these factors present in each person. Cancer patients add to this constellation of individual factors the unique case histories and conditions that each brings to the exercise equation. The material presented here is fundamental to understanding how to use exercise to minimize cancer treatment-related symptoms.

PHYSIOLOGICAL ADAPTATIONS TO EXERCISE

Exercise physiology is *the study of how structural and functional aspects of the human body are altered when exposed to acute and chronic bouts of exercise.* By measuring and monitoring changes that occur during rest and physical activity, we gain an understanding of the anatomical and physiological responses that take place during and following exercise. Since no two people respond in exactly the same way to the same exercise bout, it is important to establish baseline physiological information before initiating any exercise intervention program. Additionally, it is important to periodically reassess the individual to determine changes in these parameters and to identify when modifications in the exercise intervention should be made.

Why should cancer patients undergoing treatment or recovering from treatment consider exercise to reduce or eliminate cancer treatment-related symptoms? The answer lies in the benefits of exercise to the human system.

The more immediate changes, referred to as acute responses, serve as key indicators that help us understand the impact of exercise and measure the individual's tolerance for specific types and levels of exercise. The changes that occur over time and persist with repeated bouts of exercise are called chronic responses. These demonstrate the effects that regular exercise has on various organ systems and subsequent changes in the fitness status of the individual. The changes that can come about through exercise improve one's quality of life. The cancer patient can expect exercise to reduce or eliminate many cancer treatment-related symptoms.

Acute Physiological Responses

Table 3.1 lists the acute physiological responses to exercise. Again, acute responses are immediate changes in physiological processes that last only briefly. Once the person returns to a resting state, the changes disappear. Since no two people are exactly the same, you must consider the factors that influence these changes when predicting the degree of acute change that will occur with exercise. For the cancer patient, you need to consider factors such as stage of cancer, cancer status, type and amount of treatment, time since the last treatment, blood profile, medications, present physical fitness status, stress level, and state of restfulness. Other factors such as environmental temperature, humidity, nutrient intake, and even time of day may also contribute to the degree or type of acute changes resulting from an exercise intervention.

Cardiorespiratory System

The cardiovascular system delivers oxygen to the cells and removes metabolic waste produced by the cells. Depending on its intensity, exercise places demands on the cardiovascular system ranging from a slight rise above resting levels to increases many times above resting levels. The demand is the result of oxygen needs at the cellular level. The cells use oxygen to produce adenosine triphosphate (ATP), the basic source of energy. As exercise rises in intensity or lengthens in duration, this process must increase rapidly to supply enough energy for the work being done. Since the transport process increases due to increased demand, the heart and vascular structures must make accommodations. In this section we outline some of these accommodations and suggest how they may differ in cancer patients compared to others.

Cardiac output (CO) is the product of heart rate and stroke volume. *Heart rate* (HR) is the number of times the heart contracts and relaxes per minute, and *stroke volume* (SV) is the amount of blood pumped out of the heart by each contraction. Heart rate increases in direct proportion to the intensity of exercise. When a *submaximal exercise level* is held at a steady state, the heart rate increases for about 1 to 3 min and then plateaus until the stress of the exercise bout changes. Stroke volume is regulated by several factors. Increases or decreases in cardiac output that result from changes in stroke volume and/or heart rate are proportional to the metabolic demands of the exercise. Researchers have shown that when individuals are not highly fit, increases in cardiac output are more the result of increased heart rate than of increased stroke volume, regardless of the intensity of the exercise. The result is a greater stress on the heart because of the demand to beat faster. This can exacerbate the fatigue factors for cancer patients, whose maximal oxygen consumption is generally decreased by the side effects of cancer treatment. So fitness level can make a difference in the ability of a cancer patient to overcome treatment effects.

At rest, *peripheral blood flow* (cardiac output directed to the skeletal muscles) composes only about 21% of total blood flow compared to upward of 80% during high-intensity, exhaustive exercise (Shephard and Astrand, 1992). To meet this demand during exercise, blood flow to less active organs (kidneys, liver, digestive system) is decreased, making more blood available to the working muscles. Several mechanisms, including vasodilation and vasoconstriction (dilation and narrowing of the blood vessels, respectively), provide for this shift, making it possible for blood flow to the skeletal muscles to increase some 20 times over resting levels.

In addition to the changes in blood flow to the working muscles, blood flow to the surface of the skin increases during exercise. As the body temperature increases due to metabolic activity, a greater amount of blood is directed to the skin. Sweat glands secrete sweat to the surface and dissipate the heat brought to the periphery via the blood.

These mechanisms for regulating blood flow may be impaired in cancer patients, usually because of treatment damage to smooth muscle tissue in the vascular walls. Too much vasodilation can result in blood pooling in the extremities, causing light-headedness and a loss of balance. Failure to direct enough blood to the skin may make cancer patients especially prone to overheating. You must work carefully with each patient to be sure you are making the necessary accommodations in the patient's exercise program to avoid the consequences of exhaustion and overheating.

Table 3.1 Acute Physiological Responses to Exercise

SYSTEMS		
Cardiovascular		**Neuromuscular**
	↑ Cardiac output	↑ Sympathetic stimulation
	↑ Heart rate	↓ Parasympathetic stimulation
	↑ Contractility	
	↑ Stroke volume	
	↑ $\dot{V}O_2$	
	↑ Systolic blood pressure	
	↑ Venous return to the heart	
	Peripheral vasoconstriction	
	↓ Total peripheral resistance	
	Arteriole dilation in working muscles	
Respiratory		**Endocrine**
	↑ Minute ventilation	↑ Antidiuretic hormone
	↑ Tidal volume	↑ Aldosterone
	↑ Breathing frequency	↑ Catecholamines (epinephrine, norepinephrine)
	↑ Pulmonary blood flow	↑ Glucagon release
	↑ Dilation of bronchi	↓ Insulin release
		↑ Growth hormone
		↑ Cortisol
Metabolic		
	↓ Blood pH	
	↑ Lactic acid	
	↑ Gluconeogenesis in liver	
	↑ Glycogenolysis in liver	
	↑ Lipolysis	
	↑ Protein synthesis	
	↑ Glycolysis	

These accommodations are explored in later chapters.

Blood pressure is another cardiovascular response that may be affected by cancer. Approximately 5 L of blood in the adult human cardiovascular system must be pumped by the ventricles each minute. *Blood pressure,* or the amount of force produced by this volume of blood against the walls of all cardiovascular structures, assists in maintaining appropriate flow and distribution. The amount of pressure produced is related to blood volume, diameter of the vascular structure, and contractile force of the heart.

During exercise, an increase in blood flow is expected. During aerobic activities (lower intensity, longer duration), blood flow increase is achieved through a decrease in the resistance and a slight increase in blood pressure. During anaerobic activities (high intensity, shorter duration), blood flow is occluded due to sustained muscular contractions, and the blood pressure increases much more dramatically.

Depending on the severity and type of cancer, cancer patients can experience a decrease or an increase in their blood pressure. One of our patients who showed decreased blood pressure was severely dehydrated; if he used the rest room, his blood pressure would fall dramatically and he would have to lie down to keep from fainting. On the other hand, an increase in blood pressure could occur if there was significant resistance in any of the arteries. At the Rocky Mountain Cancer Rehabilitation Institute (RMCRI) we found this to be true in our animal research. When the aorta was exposed to a chemotherapy drug (5-fluorouracil), significant vasoconstriction occurred. Vasoconstriction increases blood pressure.

Maximal oxygen consumption ($\dot{V}O_2$max) refers to the maximal level or rate at which oxygen can be acquired, transported, and utilized by active cells during high-intensity exercise. Therefore it represents the highest rate at which oxygen can be consumed during maximal exercise performance. As the intensity or workload of an exercise bout increases, oxygen consumption increases. At the maximal level of exhaustive exercise, the oxygen consumption plateaus. Once oxygen consumption ceases to rise, $\dot{V}O_2$max has been reached. A person's age, gender, level of fitness, health status, and genetics significantly contribute to the level of maximal oxygen consumption.

Exercising at exhaustive levels is not the goal of exercise for cancer rehabilitation. However, you can use maximal oxygen consumption to prescribe exercise that will elicit specific physiological changes. Working at a given percentage of $\dot{V}O_2$max during a specific type of exercise leads to a predicted cardiac output, heart rate, stroke volume, endocrine response, muscular recruitment, neural activity, and so on. Because a given level of each of these physiological responses can be correlated with specific levels of physiological adaptation, controlling the $\dot{V}O_2$max level enables exercise specialists to achieve predicted results with their patients.

Cancer patients undergoing treatment often experience a decline in numbers of red blood cells, which are the transport vehicles for oxygen to the working muscles. Small decreases in red blood cell count can cause a decline in physical performance. Many cancer patients fatigue easily in doing simple tasks because the demand for oxygen cannot be met. Activity considered low or moderate for other individuals may be more challenging to the cancer patient. Since the effects of chemotherapy and radiation can linger for years beyond the last treatment, it is important to monitor exercise to avoid further decline in the recovery process.

Ventilation is another term for breathing. At rest, an average individual usually ventilates about 6 to 7 L of air each minute. A number of factors come into play in making this happen. Table 3.2 provides definitions of terms for the various lung volumes and capacities. An understanding of lung volumes and capacities will help when assessing lung function and designing exercise programs for cancer patients, especially lung cancer patients.

Since breathing is an autonomic function, the energy cost associated with it is often overlooked. However, the respiratory muscles must contract and relax to effect the pressure changes required to allow air to flow into the lungs and force air out of the lungs. The energy expenditure required is based on the amount of oxygen that must be allotted for these muscles. At rest, this is 2% of the total amount of oxygen being consumed. When we become active, the energy requirements of the respiratory muscles increase because ventilation increases. When exercise approaches exhaustion, the energy cost for the respiratory muscles is close to 10% of the total oxygen being consumed!

Tidal volume is the actual volume of gas breathed per ventilatory cycle. Ventilation is the product of tidal volume (TV) and breath rate or frequency (F). Generally, this volume is measured on a per minute basis *(minute ventilation)*. A

Table 3.2   Description of Lung Volumes and Capacities

Lung volumes	Description
Tidal volume (TV)	Amount of gas inspired or expired per breath
Inspiratory reserve volume (IRV)	Amount of gas that can be inspired beyond a normal inspiration
Expiratory reserve volume (ERV)	Amount of gas that can be expired beyond a normal expiration
Residual volume (RV)	Amount of gas that remains in the lungs after maximal expiration
Lung capacities	**Description**
Total lung capacity (TLC)	Amount of gas in the lungs after maximal inspiration (RV + ERV + TV + IRV = TLC)
Vital capacity (VC)	Maximal amount of gas forcefully expired after maximal inspiration (ERV + TV + IRV = VC)
Inspiratory capacity (IC)	Total amount of gas that can be inspired after a normal expiration (TV + IRV = IC)
Functional residual capacity (FRC)	Amount of gas that remains in the lungs after normal expiration (ERV + RV = FRC)

healthy average-sized adult male at rest has a tidal volume of approximately 0.5 L and a breath rate of 12 to 14. Thus, he ventilates between 6 and 7 L of air per minute. When exercise approaches the maximal level, we expect breathing rate to increase to three to four times that of rest (36 to 48 breaths/min) and tidal volume to increase to between 3 and 4 L. Ventilation at maximal exercise can reach levels near 190 L/min in a large person.

Consider a situation in which the exercise intensity increases consistently (graded exercise). From resting state to approximately 50% of $\dot{V}O_2$max, the ventilation increases linearly as the workload increases. However, above 50% of $\dot{V}O_2$max, ventilation rises at a higher rate than the workload, due more to an increase in rate of breathing than to an increase in tidal volume. The tidal volume levels off and plateaus at higher intensities of exercise. Knowing each patient's ventilatory capacity will help you understand that person's exercise tolerance. Cancer therapies, especially chemotherapy and radiation, can compromise the pulmonary system and decrease efficiency. When oxygen availability is compromised in any way, even simple physical tasks are more difficult and leave the patient feeling fatigued.

Metabolic System

With acute exercise comes an increase in the response of the metabolic system to meet the energy needs for exercise. When exercise begins, there is a lack of oxygen until the cardiovascular system can deliver oxygen to the working muscles. The muscles must use anaerobic sources such as the phosphagen system (ATP-PCr [phosphocreatine] system) and glycolysis to meet the need for ATP. *Anaerobic metabolism* decreases the blood pH due to an increase in lactic acid, the end product of glycolysis. Once steady state (oxygen supply = oxygen demand) occurs, with low-intensity exercise there is an increase in fat mobilization (lipolysis) and utilization to meet the demands for ATP for exercise. As exercise continues, protein *synthesis* is stimulated for energy use and tissue repair. For cancer patients, the exercise should be of moderate intensity to maximize the use of fats and minimize the production of lactic acid. Because of the muscle wasting that occurs with cancer treatments, it is necessary to stimulate protein synthesis so that protein is available for repair; again the exercise should not be intense or long in duration.

Neuromuscular System

Acute exercise stimulates the sympathetic nervous system and inhibits the parasympathetic nervous system, thus increasing heart rate, heart contractility, and stroke volume while decreasing total peripheral resistance and so providing the necessary transport for oxygen. The result is an improvement in blood flow and oxygen to the working muscles. At the same time, peripheral vasoconstriction occurs in the inactive tissues, forcing the blood to the working muscles. Oxygen delivery in cancer patients can be hindered with some cancer therapies. The enhanced neural stimulation can help improve blood flow and thus increase oxygen to the patient's working muscles.

Endocrine System

Acute exercise increases hormonal (endocrine) response. Levels of hormones such as antidiuretic hormone, aldosterone, epinephrine, norepinephrine, glucagon, growth hormone, and cortisol all increase with exercise. On the other hand, the level of the hormone insulin decreases. The increase in hormones help meet the demands of exercise. The decrease in insulin helps conserve blood glucose so that the brain has nutrients during exercise. The increases are transient, and values return to resting levels following exercise.

Immune System

Acute exercise-induced changes in immune system cell counts are usually variable. Some studies have shown changes in immune function, but usually within the context of intense acute exercise. Many of the changes are too transient to affect immune defenses. Once again, for cancer patients, moderate low-intensity, short-duration exercise can produce a positive response, especially if the patient participates in exercise training.

Chronic Responses to Exercise

The changes just discussed occur in direct response to exercise, and values soon return to normal. However, with regular performance of exercise, certain physiological changes persist. These chronic adaptations mean improvements in the cardiovascular, pulmonary, metabolic, skeletal muscle, endocrine, and immune systems, which will improve patients' functional capacity. Regular exercise can also have psychosocial benefits that are important for cancer survivors.

Chronic adaptations occur at different rates for different individuals. Each system responds differently to a continuous dose of exercise. For example, the cardiovascular and metabolic systems show rapid improvements in apparently healthy individuals of low to moderate fitness level when they perform exercise for 40 to 60 min, six days per week, at intensities ranging between 70% and 90% of $\dot{V}O_2$max. Within approximately one to three weeks, such individuals can improve $\dot{V}O_2$max and ventilation, decrease submaximal heart rate for given workloads, and perform work at higher intensities for longer periods of time (Hickson, 1981).

The decline in specific physiological systems brought on by cancer tumors or cancer therapies is likely to delay the chronic response to exercise in cancer patients as compared with healthy individuals. Additionally, cancer patients may not be able to tolerate levels of exercise that bring about rapid changes. Thus, a slower progression toward desired adaptations is generally necessary. Chapters 6 and 7 cover the development and monitoring of individualized exercise programs to effect specific physiological changes for cancer patients.

Cardiovascular Adaptations

The effects of long-term endurance exercise intervention can be seen both in the resting state and during physical activity (table 3.3). In general, the heart muscle and associated vascular structure become stronger, improving the capacity to accept and deliver blood. For example, a regimen of prescriptive aerobic exercise can decrease resting heart rate as much as 10 to 15 bpm (Frick, Elovainio, and Somer, 1967). The heart muscle itself can increase in size; however, this effect usually occurs after many months of consistent endurance exercise intervention. The increase in size results from changes in the wall thickness of the ventricles with concomitant increase in cavity size (Smith and Mitchell, 1993). The increase in size allows the heart to contract with greater strength and to accommodate a larger end-diastolic volume. An increased end-diastolic volume means an increase in stroke volume. With a larger stroke volume, the heart does not have to beat as often to produce the necessary cardiac output. Although maximum cardiac output can increase with chronic endurance exercise, at rest there is little change. In addition to these alterations, individuals who consistently

Table 3.3 Chronic Physiological Responses to Exercise

SYSTEMS	
Cardiovascular	**Neuromuscular**
Left ventricular hypertrophy	↑ Number and size of mitochondria
↑ Blood volume	↑ Krebs cycle and ETC enzyme activity
↑ Left ventricular compliance	↑ Krebs cycle and ETC enzyme concentration
↑ End-diastolic volume	↑ Skeletal muscle myoglobin
↑ Stroke volume	↑ Capillary density
↓ Resting heart rate	↑ Protein synthesis
↑ Maximum cardiac output	Skeletal muscle hypertrophy
↑ Red blood cells	↓ Muscular fatigue
↑ Hemoglobin	↑ Muscular force production
↑ Ejection fraction	
Respiratory	**Endocrine**
↑ Respiratory muscular endurance	↓ Release of catecholamines
Higher minute ventilation at maximal exercise	↓ Growth hormone response
↑ Ability to remove CO_2	↓ Cortisol response
↑ $\dot{V}O_2max$	
Metabolic	
Delayed onset of blood lactic acid	
↑ Lactic acid oxidation	
↑ Glycogen storage	
↑ Hexokinase activity	
↓ Glycogen depletion	
↓ LDH activity (less lactate)	
↑ Mobilization and oxidation of fat	
↑ Hormone-sensitive lipase activity	

Note: LDH = lactate dehydrogenase; ETC = electron transport chain.

perform endurance-type activities may also experience slight decreases in systolic and diastolic blood pressures and increases in blood volume and red blood cell production, with concomitant increases in hemoglobin concentration. These changes mean less stress on the heart and a conservation of energy, leading to a lessening of fatigue in cancer patients.

Since endurance-type activity forces the heart to accept greater blood volumes during exercise, and since muscle-strengthening activities force the heart to work against greater loads, endurance exercise is most appropriate for increasing the efficiency of the cardiovascular system.

When comparing the negative effects of cancer therapies on the cardiovascular system with the positive effects of exercise, you can see that all the adaptations that occur with chronic exercise can minimize the toxic treatment effects. The cancer patient will have lower resting heart rate, greater stroke volume, greater cardiac output, increased blood volume, and increased hemoglobin concentration—all of which will improve the patient's oxygen delivery to the working muscles. This increases the patient's functional capacity, making daily tasks more manageable. At RMCRI, we are definitely seeing these improvements in our cancer patients.

Pulmonary Adaptations

Although there is little evidence to suggest that chronic exercise causes drastic changes in normal pulmonary function, people who maintain consistent endurance exercise intervention programs seem to have larger lung volumes and increased diffusion capacity at rest and during exercise. Moreover, endurance exercise produces greater metabolic efficiency, so the oxygen demand is lessened at rest and during submaximal exercise. The cardiovascular and metabolic systems seem to adapt to exercise intervention more than the pulmonary system.

Cancer patients who have lung involvement due to disease, metastasis, or treatment may experience greater benefits from endurance exercise than normal healthy individuals (Schneider, Dennehy, Roozeboom, and Carter, 2002; Schneider, Bentz, and Carter, 2002b). Patients presenting with chronic obstructive pulmonary disease (COPD), for example, see increased exercise capacity, increased pulmonary function, and improved quality of life (Maltais, Simard, Jobin, Desgagnes, and Le Blanc, 1996). The aim of the exercise prescrip-

tion for these individuals should be to decrease dyspnea (labored breathing) and improve exercise endurance.

Metabolic System Adaptations

Improvements in the metabolic system occur following chronic exercise training. Physiological adaptations take place in the aerobic and anaerobic energy systems. Changes include a delay in the onset of blood lactic acid, increased lactic acid oxidation, increased glycogen storage, decreased glycogen depletion, and increased mobilization and utilization of fats for energy production.

Because of the negative effects of treatments, cancer patients have to work harder and expend more energy than others to do even simple physical tasks. They therefore have high levels of lactic acid because they use the anaerobic metabolic system for even submaximal work. High lactic acid levels cause fatigue. As already mentioned, training can delay the onset of lactic acid and enhance the use of the aerobic system (fat). This can spare glycogen, reduce lactic acid, decrease fat stores, and reduce fatigue.

Skeletal Muscle Cell Responses

Chronic endurance exercise intervention improves the efficiency of skeletal muscle cells to perform contractions. The following are some of the cellular adaptations that relate directly to this improved efficiency:

- The mitochondria, where the oxidative processes for energy metabolism occur, increase in number and size.
- An increase in myoglobin (a muscle protein responsible for transporting oxygen to the mitochondria) occurs.
- The concentration of enzymes needed to catalyze reactions during the oxidation of glucose rises.
- Muscle hypertrophy (size) increases through increased protein synthesis.
- Muscle cells increase in their capacity to store glycogen.
- Muscle cells become more efficient in their ability to utilize fats as fuel for contractions.

The improvements in the muscular system help cancer patients overcome the debilitating muscular weakness that they experience as a consequence of muscle wasting from treatment toxicities. Exercise increases the size of the muscle through increased protein synthesis and decreased protein degradation. Force production (strength) is enhanced. We have found at RMCRI that patients' debilitating fatigue is more closely associated with muscular weakness than it is with the reduction in functional capacity. Therefore a strength component should definitely be a part of cancer rehabilitation.

Endocrine System Responses

The hormonal response to exercise—among hormones such as growth hormone, cortisol, epinephrine, and norepinephrine—is blunted with chronic training. This means that at a given workload, the hormonal response is less after exercise training than prior to training. For cancer patients, then, daily work activities are easier after training.

Exercise and Natural Immunity

Chronic intense endurance exercise has been associated with an increased risk for infectious illness. This finding (Nieman, Johanssen, Lee, and Arabatzis, 1990; Peters, Goetzsche, Joseph, and Noakes, 1996) has been consistent throughout the literature. In contrast, susceptibility to infection was not altered in shorter, less competitive events. Castell, Newsholme, and Poortmans (1996) reported that the incidence of upper respiratory tract infection was 50% lower during the week after competition in middle distance runners compared with long distance runners. These studies support the importance of using exercise of moderate intensity and duration for cancer patients.

Nieman et al. (1995) investigated the effects of moderate exercise on 16 previously diagnosed (3 years since diagnosis) breast cancer patients who underwent surgery, chemotherapy, and/or radiation treatments. Women were assigned to exercise or nonexercise groups. The exercise group participated in an eight-week, three times per week, 60-min weight-training and moderate aerobic activity program. Pre- and postexercise assessments were carried out for aerobic performance, leg strength, and concentrations of circulating lymphocyte subsets and natural killer cytotoxic activity (NKCA). The exercise group improved in aerobic capacity and strength. However, NKCA and concentrations of circulating T cells and natural killer cells were not altered compared to values in the nonexercise group. The researchers concluded that moderate ex-

ercise has no significant effect on the function of natural killer cells in breast cancer patients and thus does not compromise natural immunity. Peters, Lotzerich, Niemeier, Schule, and Uhlenbruck (1994) obtained similar results in stage I and II breast cancer patients (6 months postsurgery). Natural immunity was not compromised. In fact, the patients had enhanced cellular immune function after a six-month cycle ergometry exercise program. Nieman (1994) reported that moderate exercise enhanced the immune system and that heavy intense exercise suppressed the immune system. The mechanism by which exercise enhances the immune system is not clearly understood. Some possible explanations are that

- the increase in body temperature creates an unfavorable environment for invading pathogens;
- exercise may lead to a more favorable balance between the immune system, the release of hormones, and the body's response to stress;
- exercise-induced reductions in body fat may provide a healthier body to combat infections; and
- exercise may lead to alterations in the components of the immune system itself.

Body Composition

As mentioned earlier, exercise—especially low-intensity, long-duration exercise—utilizes fat metabolism. The increase in fat metabolism reduces fat mass. Exercise also increases muscle mass or lean body mass. This reduction in body fat enhances one's health status. Cancer patients often gain weight or lose extreme amounts of weight during cancer and cancer treatments. Exercise helps patients lose weight if they need to do so; or, utilizing strength training, patients can gain muscle mass.

Psychological Effects

Regular physical exercise improves mood, body image, self-concept, and sleep patterns. Exercise can decrease anxiety, tension, anger, depression, and feelings of helplessness in cancer patients. Research has shown that breast cancer patients who exercise have a significantly better quality of life than breast cancer patients who do not exercise (McTiernan, Gralow, and Talbott, 2000).

FURTHER READINGS

Bowers, R.W., and E.L. Fox. 1992. *Sports physiology* (3rd ed.). Dubuque, IA: Brown. Comment: This is a very basic text on the influence of exercise on the various physiological systems. It provides basic information on the effects of training.

Brooks, G.A., T.D. Fahey, and T.P. White. 2000. *Exercise physiology: Human bioenergetics and its applications* (3rd ed.). Mountain View, CA: Mayfield. Comment: This text contains graduate-level information on the influence of exercise on physiological systems, as well as current research in the area of exercise physiology.

DeVries, H.A., and T.J. Housh. 1994. *Physiology of exercise* (5th ed.). Dubuque, IA: Brown & Benchmark. Comment: This undergraduate text provides information on the influence of exercise on the body. The text is somewhat hard to understand.

Foss, M.L., and S.J. Keteyian. 1998. *Fox's physiological basis for exercise and sport* (6th ed.). Boston: McGraw-Hill. Comment: This book, intended as a senior-level undergraduate text, offers information on the influence of exercise and training on the physiological systems and includes application of the various concepts in exercise physiology. This is an excellent text on exercise.

Summarizing the Positive Effects of Exercise on Cancer Patients

As noted in chapter 2, all cancers and cancer treatments cause alterations centrally as well as in the periphery of the cardiovascular system. These alterations result in ventricular dysfunction, a narrowing of the coronary arteries, and a reduction in hematopoietic cells within the bone marrow, thus affecting the composition of the peripheral blood. As we have seen in this chapter, exercise reduces the stress on the heart and blood vessels, enhancing the ability of the heart and lungs to deliver oxygen efficiently to the working tissues in spite of cancer and cancer treatment damage. The evidence obtained at RMCRI (Dennehy, Carter, and Schneider, 2000a, 2000b; Dennehy, Schneider, Carter, Bentz, Stephens, and Quick, 2000; Dennehy, Bentz, Stephens, Carter, and Schneider, 2000; Dennehy, Bentz, Fox, Carter, and

Schneider, 1998; Schneider et al., 2000, 2002; Schneider, Bentz, and Carter, 2002a, 2002b; Stephens et al., 2000) demonstrates significant improvements in functional capacity, resting heart rate, time on treadmill, forced vital capacity, and range of motion, with concomitant reductions in perception of fatigue and depression, following six-month exercise interventions in cancer patients compared to nonexercise cancer patient controls. Evidence from these studies showed that muscular strength and endurance significantly improved in cancer patients. In fact, the muscular weakness observed in the cancer patients appeared to be more highly correlated to the debilitating fatigue that patients experienced than to cardiovascular endurance.

Exercise has consistently improved the quality of life for cancer patients both physically and emotionally. The key is to ensure that exercise is at moderate intensity. The principle of individuality dictates that you must tailor the exercise prescription specifically to each person, especially for someone with compromised physiological health such as a cancer patient.

BASIC PRINCIPLES OF EXERCISE PRESCRIPTION

If one is to expect specific outcomes from therapeutic exercise intervention, it is important to apply the fundamental principles of exercise. These principles serve as the basis for developing and conducting an exercise intervention. Although the principles underlie all levels and types of exercise, additional considerations may be necessary in work with cancer patients. Throughout this section we discuss these special considerations along with the basic principles.

Principle of Specificity

The adaptations or responses to any exercise intervention are highly specific. This suggests that prescribed exercise will result in a specific and predictable outcome. For example, if the goal is to decrease body fat, then the exercise performed must be designed to use the energy systems in a specific way—to use fat as the main fuel. If the goal is to increase muscular endurance of the legs, then the work must consist of activities that train the muscle fibers of the legs to be more efficient.

To apply this principle, you must take into account the energy systems used to fuel the work, the muscle group and motor pattern involved during the work, and the mode or means by which the work will be carried out. By following the *principle of specificity* and carefully considering each of these elements, you can target precise outcomes for a patient and design a protocol maximizing the benefits of each exercise session.

Specificity of Energy Systems

The specificity of energy systems implies that exercises are not all fueled in the same way. The cells of the human body require a particular molecule *(adenosine triphosphate or ATP)* to fuel work. Adenosine triphosphate includes several high-energy phosphate bonds (figure 3.1). The body obtains energy for exercise by breaking the third phosphate bond, releasing the energy stored there. The body uses three different chemical pathways to manufacture ATP: the ATP-PCr system, the glycolytic system, and the oxidative system. The source of the energy and raw materials needed for ATP creation is foods composed of energy-yielding nutrients such as carbohydrates, proteins, and fats.

The *ATP-PCr system* is a relatively simple process of resynthesizing ATP. Phosphocreatine releases energy when an enzyme (creatine kinase) causes it to split into creatine and phosphate. This energy is then used to couple the phosphate onto adenosine diphosphate (ADP) to form ATP.

The *glycolytic system,* which involves the release of energy through the degradation of glucose (known as glycolysis), is more complex than the ATP-PCr system. In this system, 12 enzymatic reactions must occur in sequence to release 2 mol of ATP from each mole of glucose. The first stage of the system involves converting glucose to glucose-6-phosphate, a process that requires using some of the ATP that is already available. Glycogen can also be used to fuel the glycolytic system. In this case, 3 mol of ATP are released from each mole of glycogen because glycogen does not require any available ATP to be converted to glucose-6-phosphate. The end product of glycolysis is lactic acid that accumulates rapidly in the muscles and blood, inhibiting further breakdown of glucose and severely limiting the subsequent production of ATP. Without sufficient fuel, the working muscle cells can no longer perform contractions.

The ATP-PCr and glycolytic systems are anaerobic—that is, they do not require oxygen—and together they are the primary means of fueling muscle work during the first few minutes of high-intensity exercise. However, since neither of these systems produces high amounts of energy, they are not able to supply sufficient energy for vigorous exercise for very long periods of time. In fact, if no other energy system were available to assist in the production of ATP, intense exercise would be limited to 2 to 3 minutes!

The *oxidative system* is by far the most complex of the energy systems. As the name implies, oxygen is used to break down energy-yielding nutrients and convert some of the released energy to ATP. Thus, the process is sometimes referred to as *aerobic metabolism.* The oxidative system can provide ample amounts of ATP for longer-duration, less intense work as long as fuel and oxygen are made available to the muscle cells. The aerobic process can produce up to 38 molecules of ATP from a single molecule of glucose. This is 19 times more than the yield from glycolysis!

As a cancer exercise specialist, you should be knowledgeable about how each energy system

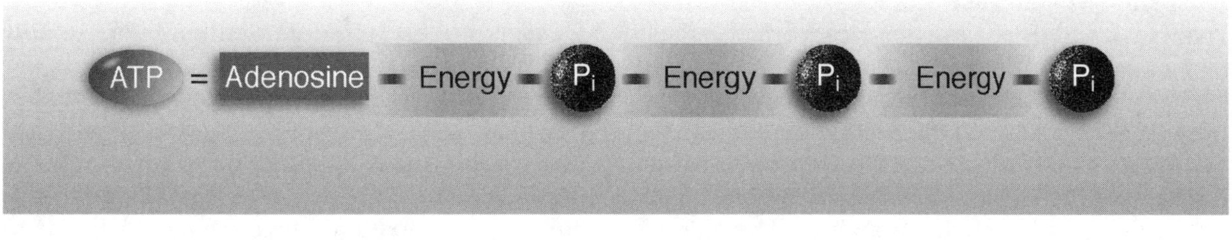

Figure 3.1   The structure of an adenosine triphosphate (ATP) molecule, showing the high-energy phosphate bonds.
Reprinted, by permission, from J.H. Wilmore and D.L. Costill. 1999. *Physiology of Sport and Exercise,* 2nd ed. (Champaign, IL: Human Kinetics), 120.

contributes to an exercise bout. Since many cancer patients experience a decline in energy level during and following treatment, you need to design prescriptive exercise to maximize available energy and enhance energy production. In addition, a regular exercise program can produce more efficiency in the use of energy, helping to conserve energy stores.

Specificity of Motor Patterns

The specificity of muscle groups or motor patterns implies that a movement must be practiced exactly and consistently in order to be perfected. Movements should be precise if the goal is to perform efficiently and productively. Cancer exercise specialists, in conjunction with the physical therapist or medical director, often have the task of helping postsurgical patients relearn movement patterns. Motor patterns require the use of specific muscles that function at a particular speed through a given range of motion. For example, walking requires that one foot be placed in front of the other at a given rate. Performing this action too slowly may result in a loss of balance.

Or consider what type of exercise intervention might be necessary to produce efficient movement following a radical mastectomy. In this type of surgery, the chest wall muscles are altered and axillary lymph nodes are removed, resulting in limited movement of the arm and weakness through the shoulders and chest. You must know how to work with patients to produce efficient movements given the existing musculature and potential limit in lymphatic drainage. Thus you must be able to analyze movement limitations and select exercise interventions that will produce the desired outcome.

Specificity of Modality

The specificity of modality in many ways is directly associated with the specificity of muscle groups and motor patterns. To select the correct mode of exercise to facilitate desired outcomes, you must know which muscles are used to perform a given task and how these muscles work together and alone to contribute to the movement desired. The mode should allow the patient to mimic the desired performance. If the desired outcome is to climb stairs more efficiently, it is important to incorporate practice on stairs. Swimming, although it may help to increase endurance, will help little with efficient stair climbing.

You must also be concerned about modality for cancer patients who are limited in their ability to perform weight-bearing activities. For example, some patients develop edema in the lower extremities during treatment and are unable to walk any significant distance. In these cases, it is necessary to use other types of exercise that yield the same outcome. A patient who cannot walk because of lower leg swelling may be able to ride a bike since the bike will carry the weight. The patient can still obtain improvements in functional capacity.

Principle of Overload

Positive changes in muscular performance (strength and endurance), cardiovascular efficiency, flexibility, and range of motion can occur if exercises are carried out at a higher level than is normally required. This concept is the basis for the *principle of overload*. Strength and endurance, flexibility, and range of motion increase only when muscles are made to perform greater-than-normal levels of work.

To ensure that the cancer patient will progress without injury or undesired stress and will have the best opportunity to reach specified goals, you must develop exercise prescriptions that define *how* the work should be administered. In other words, there are several ways to accomplish overload, and each patient's needs should dictate exactly which methods are most appropriate. Manipulating the physical work components of frequency, intensity, and duration will alter exercise levels. When determining how best to use these factors to enhance strength and endurance for patients, you must take into account the degree of fatigue the individual is experiencing, present dietary restrictions or limitations, treatment regimen, blood profile, medications, and any other specific conditions that might limit participation and confound the effects of the exercise intervention.

Frequency

Frequency of exercise intervention refers to how many exercise sessions a person performs per week during an exercise program. The number depends on the status of the cancer patient. Normally, if the goal is to improve cardiovascular endurance and maintain age- and gender-appropriate body composition, the recommendation is to exercise at least three times per week with no more

It is vital that you consult frequently with every patient. Listen more than you talk. Ask questions and listen to the answers to be sure you understand what is going on with treatments and the patient's reactions to them; and that you know what effect the patient's exercise is having at all times. Otherwise, you cannot adjust prescriptions appropriately as symptoms change.

than two days between exercise interventions. But cancer patients often cannot exercise at a high enough intensity or for a long enough duration to make progress at that rate, so it is often necessary to adjust all three elements (frequency, intensity, and duration) in unusual ways to achieve the desired results. According to the American College of Sports Medicine (ACSM), for example, patients exhibiting very low functional capacity (<3 METs [metabolic equivalents]) can benefit from several short exercise sessions of 3 to 5 min conducted throughout a single day. As the patient progresses (~5 METs), exercise session time can be increased and the frequency per day can be reduced. Continued progress (>5 METs) means that the patient may advance to longer single sessions (20-30 min), three to five days per week (American College of Sports Medicine, 2000).

Intensity

Intensity of exercise refers to the level or degree of the work being performed. Because an individual's heart rate increases in a linear fashion as the intensity of exercise increases, we can measure this factor by noting the patient's heart rate maximum (HRmax). The maximum heart rate is defined as the number of beats per minute exhibited at the point of exhaustion.

Just as heart rate increases linearly as the intensity of exercise increases, so does oxygen uptake. Thus, oxygen uptake can also indicate exercise intensity. The oxygen uptake measured at the point of exhaustion is referred to as $\dot{V}O_2$max. $\dot{V}O_2$ is measured in L/min or in ml $\cdot$ kg^{-1} $\cdot$ min^{-1}, and is determined through analysis of expired gases of the exercising individual. The equipment required to make these measurements is rather expensive and complex, so determining $\dot{V}O_2$max is usually not a practical option for measuring exercise intensity for cancer patients.

To determine the appropriate exercise intensity for a patient, you must first know the individual's HRmax. The most accurate way to establish this is to have the patient undergo a graded exercise test (explained in chapter 5). Another common, though less accurate, way to determine HRmax is to subtract the patient's age from 220. (If, for example, you are 56 years old, your HRmax is approximately 220 – 56, or 164 bpm.) The American College of Sports Medicine recommends that normal healthy individuals exercise at an intensity level equal to between 50% and 85% of their heart rate reserve (HRR), which is determined by subtracting the resting heart rate value from the maximum heart rate. The percentage of HRR is then added to the patient's resting HR to determine target HR. (To continue the example just cited, given an HRmax of 164 and a resting heart rate of 68 bpm, your HRR will be 164 – 68, or 96 bpm. You want the patient to work at 65% of HRR; this equals 62.4 bpm. Then add the patient's RHR; this gives a target heart rate of 130 bpm. So, you would work the patient at an intensity of 130 bpm.) The most common method of determining the intensity of exercise for each patient is to use the Karvonen formula (figure 3.2), which is

$$\text{target heart rate} = \text{resting heart rate} + [\% \text{ exercise intensity} \times \text{HRR}]$$

For healthy individuals, as just noted, ACSM recommends intensity levels between 50% and 85% of HRR. For cancer patients, however, this range may need to be lower. The percentage to use in the Karvonen formula depends on how long the patient wishes to exercise (a higher intensity will require shorter duration); whether the patient needs a lower intensity to make it possible (or simply easier) to maintain the workload; how fast you and your patient wish improvement to occur; and the health status of the patient.

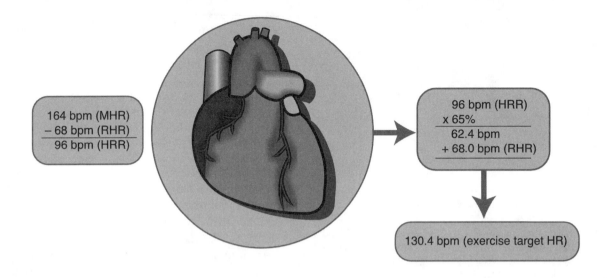

A 56-year-old female has a resting heart rate of 68 bpm. Her maximum heart rate can be estimated by subtracting her age from 220 (220 − 56 = 164 bpm). She wants to limit her exercise intensity to 65% of her HPR.

164 bpm (MHR)
− 68 bpm (RHR)
96 bpm (HRR)

96 bpm (HRR)
x 65%
62.4 bpm
+ 68.0 bpm (RHR)

130.4 bpm (exercise target HR)

This woman should maintain her heart rate at 130 bpm during her exercise session.

Figure 3.2   The Karvonen formula. MHR = maximum heart rate (HRmax); RHR = resting heart rate; HRR = heart rate reserve.

As is clear from the Karvonen formula, patients have a lower maximal heart rate as they age. Thus, even low-intensity physical activity can elevate the heart rate and increase cardiac output and can exacerbate fatigue rather than lessen it. For this reason you need to be particularly careful not to prescribe levels of exercise that are too intense for older patients. Additionally, if patients are on medications that affect the heart rate, they can use ratings of perceived exertion—although cancer patients usually underestimate their effort because exercising is so much easier than going through cancer and cancer treatments.

Duration

Duration of exercise refers to the amount of time work is performed during each exercise session. The criteria for selecting the appropriate duration of an activity are the present fitness and health status of the individual, the intensity of the work being performed, and the impact of the previous exercise session. For a normal healthy population, ACSM (2000) recommends that moderate-intensity

work (>45% $\dot{V}O_2$max) be sustained for between 20 and 30 min to provide benefits in an appropriate time course. But for a cancer patient, you need to adjust these parameters, taking the individual's functional capacity into account. Remember—use low intensity, short duration to start, and progress from there.

Principles of Progression, Adaptation, and Individuality

The other principles you must have in mind as you design exercise programs for cancer patients are those of progression, adaptation, and individuality.

• **Progression.** For your patients to experience improvements in their fitness levels, you must continually adjust the frequency, intensity, and/ or duration of each person's exercise regimen to provide the necessary overload. This process of continual adjustment is guided by the *principle of progression.* If progression is too rapid for cancer patients, the result will be greater stress on the body and increased symptoms of fatigue. If progression occurs too slowly, the benefits of the exercise intervention will be less than expected, and a loss of interest and participation may be the result. You must monitor progression closely so that your patients can realize maximum benefits while avoiding injury or setbacks.

• **Adaptation.** According to the *adaptations principle,* adaptations occur as an individual participates in prescriptive exercise designed to produce specific changes in physiological response to stressors. The changes are subtle, yet predictable. The cardiovascular system, muscles, bones, immune system, and blood profile all benefit from appropriate and consistent exercise. Because cancer patients normally experience such drastic reductions in their quality of life, it is vital that they exercise regularly to achieve the positive adaptations that will make them feel—and probably heal—better.

• **Individuality.** No two patients have exactly the same exercise requirements. Cancer patients differ in their cardiovascular, pulmonary, and muscular strength and endurance; each one has had a unique history and personality that have affected that individual's motivation; and each has unique personal and physical needs. Thus you must recognize that a "one size fits all" exercise program will not benefit every patient. Maximum benefits come from a prescriptive exercise intervention that is carefully crafted to "match" each patient and that is closely monitored and modified as the patient's needs and desires change. Therefore you must always keep the *individuality principle* in mind as you prescribe exercise for cancer patients.

Cancer patients undergoing treatment are especially subject to physical and psychological change as they respond to their particular therapeutic regimens. It is essential for you to be alert to these changes and to know how to vary the exercise intervention to ensure that exercise does not further debilitate the patient. Examples of specific cases in later chapters illustrate current understanding of how to gauge this. The Rocky Mountain Cancer Rehabilitation Institute is completing research to provide more specific data than are now available to help cancer exercise specialists work more effectively with patients who have various types and stages of cancer (Schneider et al., 2002, 2002b).

SUMMARY

Understanding the fundamentals of exercise is a key to understanding how physical activity can ameliorate many cancer treatment-related symptoms. Exercise brings about immediate, or acute, physiological responses, as well as longer-term, or chronic, responses. The acute responses in the cardiorespiratory, metabolic, neuromuscular, endocrine, and immune systems serve as key indicators that help us understand the impact of exercise and measure the individual's tolerance for specific types and levels of exercise.

Whereas the acute physiological changes are short-lived, with values returning to normal soon after the exercise bout has ended, chronic responses persist over time, leading to longer-term physiological adaptations as well as to improvements in body composition and in psychological factors such as mood, body image, self-concept, and sleep patterns. In the cardiovascular system, for example, the adaptations that occur with chronic exercise correspond to and minimize the negative effects of cancer therapies, with patients improving their functional capacity and resting heart rate, among other cardiovascular parameters. Chronic exercise also leads to improvements in range of motion and muscular strength and endurance, as well as to psychological benefits and a lessening of whole-body fatigue.

For cancer patients to achieve these specific outcomes through exercise, the cancer exercise specialist must design each person's program according to the fundamental principles of exercise and also in view of each patient's particular needs, limitations, capacities, case history, and cancer-related conditions. The principle of specificity states that one can target precise outcomes by selecting the energy systems used to fuel the exercise, the muscle group or motor pattern involved, and the mode or means by which the exercise will be performed. The principle of overload states that strength and endurance, flexibility, and range of motion increase only when muscles are made to perform greater-than-normal levels of work. The cancer exercise specialist manipulates overload by altering the frequency, intensity, and duration of exercise given the patient's particular situation, needs and desires, and physical status. Other exercise principles are progression (continual adjustment of the program to achieve maximal benefit while avoiding setbacks), adaptations (subtle yet predictable changes one can expect from the program), and individuality. Cancer exercise specialists need to match all components of a patient's exercise program individually to that patient's limitations and goals.

STUDY QUESTIONS

1. Differentiate between acute and chronic responses to exercise within each physiological system.

2. Explain the adaptations that occur within each system with chronic exercise.

3. What are the principles of exercise most specific to prescribing exercise interventions for cancer patients?

4. What components of exercise would you manipulate to produce an overload? How would you accomplish this with cancer patients?

REFERENCES

American College of Sports Medicine. 2000. *ACSM's guidelines for exercise testing and prescription* (6th ed.), ed. B.A. Franklin. Baltimore: Williams & Wilkins.

Bowers, R.W., and E.L. Fox. 1992. *Sports physiology* (3rd ed.). Dubuque, IA: Brown.

Brooks, G.A., T.D. Fahey, and T.P. White. 2000. *Exercise physiology: Human bioenergetics and its applications* (3rd ed.). Mountain View, CA: Mayfield.

Castell, L.M., E.A. Newsholme, and J.R. Poortmans. 1996. Does glutamine have a role in reducing infections in athletes? *Eur J Appl Physiol* 73:488–490.

Dennehy, C.A., A. Bentz, K. Fox, S.D. Carter, and C.M. Schneider. 1998. Breast cancer risk profile of sedentary, active, and highly trained women. *Med Sci Sports Exerc* 30(5):S63.

Dennehy, C.A., A. Bentz, K. Stephens, S.D. Carter, and C.M. Schneider. 2000. The effect of cancer treatment on fatigue indices. *Med Sci Sports Exerc* 32(5):S234.

Dennehy, C.A., S.D. Carter, and C.M. Schneider. 2000a. Physiological manifestations of prescriptive exercise on cancer treatment-related fatigue. *Physiologist* 43(4):357.

Dennehy, C.A., S.D. Carter, and C.M. Schneider. 2000b. Prescriptive exercise intervention for cancer treatment-related fatigue. *Physical Activity and Cancer: Cooper Aerobics Institute,* No. 4, p. 22.

Dennehy, C.A., C.M. Schneider, S.D. Carter, A. Bentz, K. Stephens, and K. Quick. 2000. Exercise intervention for cancer-related fatigue. *Res Q Exerc Sport* 71(1, suppl):A-27.

DeVries, H.A., and T.J. Housh. 1994. *Physiology of exercise* (5th ed.). Dubuque, IA: Brown & Benchmark.

Foss, M.L., and S.J. Keteyian. 1998. *Fox's physiological basis for exercise and sport* (6th ed.). Boston: McGraw-Hill.

Frick, M., R. Elovainio, and T. Somer. 1967. The mechanism of bradycardia evoked by physical training. *Cardiologia* 51:46–54.

Hickson, R.C. 1981. Time course of the adaptive responses of aerobic power and heart rate to training. *Med Sci Sports Exerc* 13:17–20.

Maltais, F., C. Simard, J. Jobin, P. Desgagnes, and P. Le Blanc. 1996. Oxidative capacity of the skeletal muscle and lactic acid kinetics during exercise in normal subjects and in patients with COPD. *Am J Resp Crit Care Med* 153:288–293.

McTiernan, A., J. Gralow, and L. Talbott. 2000. *Breast fitness.* New York: St. Martin's Press.

Nieman, D.C. 1994. Exercise, upper respiratory tract infection, and the immune system. *Med Sci Sports Exerc* 26(2):128–139.

Nieman, D.C., V.D. Cook, D.A. Henson, J. Suttles, W.J. Rejeski, P.M. Ribisl, O.R. Fagoaga, and S.L. Nehlsen-Cannarella. 1995. Moderate exercise training and natural killer cell cytotoxic activity in breast cancer patients. *Int J Sports Med* 16(5):334–337.

Nieman, D.C., L.M. Johanssen, J.W. Lee, and K. Arabatzis. 1990. Infectious episodes in runners before and after the Los Angeles Marathon. *J Sports Med Phys Fit* 30:316–328.

Peters, C., H. Lotzerich, B. Niemeier, K. Schule, and G. Uhlenbruck. 1994. Influence of moderate exercise training on natural killer cytotoxicity and personality traits in cancer patients. *Anticancer Res* 14:1033–1036.

Peters, E.M., J.M. Goetzsche, L.E. Joseph, and T.D. Noakes. 1996. Vitamin C as effective as combinations of anti-oxidant nutrients in reducing symptoms of upper respiratory tract infection in ultramarathon runners. *So African Sportsmed* 3:23–27.

Schneider, C.M., A. Bentz, and S.D. Carter. 2002a. *The influence of prescriptive exercise rehabilitation on fatigue indices.* Manuscript in preparation.

Schneider, C.M., A. Bentz, and S.D. Carter. 2002b. *Prescriptive exercise rehabilitation adaptations in cancer patients.* Manuscript in preparation.

Schneider, C.M., C.A. Dennehy, M. Roozeboom, and S.D. Carter. 2002. A model program; exercise intervention for cancer rehabilitation. *J Integr Cancer Ther* 1(1):76–82.

Schneider, C.M., K. Stephens, A. Bentz, K. Quick, S.D. Carter, and C.A. Dennehy. 2000. Prescriptive exercise rehabilitation adaptations in cancer patients. *Med Sci Sports Exerc* 32(5):S234.

Shephard, R.J., and P-O. Astrand. 1992. *Endurance in sport: The encyclopedia of sports medicine.* Vol. II. Oxford: Blackwell Scientific.

Smith, M.I., and J.H. Mitchell. 1993. Cardiorespiratory adaptations to exercise training. In *Resource manual for guidelines for exercise testing and prescription,* ed. J.L. Durstine (2nd ed.). Baltimore: Williams & Wilkins.

Stephens, K., C.M. Schneider, A. Bentz, M. Lapp, K. Quick, S.D. Carter, and C.A. Dennehy. 2000. The influence of time from cancer treatment on selected fatigue indicators. *Med Sci Sports Exerc* 32(5):S233.

4

Research and Basic Guidelines for Cancer Exercise Rehabilitation

Patients recovering from cancer and cancer treatments incur debilitating side effects as detailed in chapter 2. Exercise interventions can aid cancer patients during treatment and recovery. In this chapter we first review data from research on the benefits of exercise for cancer patients and then discuss research relating to the development of exercise guidelines for these patients. We next present some general guidelines based on research and finally describe the guidelines thus far developed at the Rocky Mountain Cancer Rehabilitation Institute (RMCRI) for implementing exercise interventions for cancer patients.

Cancer-related fatigue and other treatment-related symptoms that occur in a majority of cancer patients are severe and activity limiting, and they can last for months or even years after treatment. Cancer patients who experience fatigue and muscle weakness are usually told by their health professionals to "get more rest." Winningham (1992) reported that excessive rest for cancer patients actually results in increased fatigue and physiological deterioration. Evidence of beneficial

exercise intervention effects for cancer survivors is accumulating.

STUDIES ON EXERCISE BENEFITS FOR CANCER PATIENTS

Researchers are increasingly reporting the benefits of exercise for cancer patients. For example, studies have shown that exercise leads to improvements in cancer patients' aerobic capacity and muscular strength. Exercise also significantly affects the fatigue that so many cancer patients experience, alleviates other symptoms such as nausea, and helps improve quality of life.

Aerobic Capacity

As early as 1981, benefits of exercise for cancer patients were reported. Buettner and Gavron (1981) found that patients with a history of cancer

"I feel much better after an hour of guided exercise activity in RMCRI. When I finish an exercise session, my energy level is 100% better."

who completed an eight-week training program showed significant improvements in aerobic capacity compared with a sedentary group of cancer patients. The majority of these patients had completed cancer treatment. Winningham and MacVicar (1985), on the other hand, investigated the effects of an exercise program (10 weeks, three times weekly, on a cycle ergometer) on women with stage II breast cancer who were on chemotherapy. The exercising cancer patients improved 20.7% on aerobic capacity while the nonexercising cancer patients showed a decreased aerobic capacity (–1.8%). Later, MacVicar, Winningham, and Nickel (1989) completed an exercise (10 weeks, three times weekly, on a cycle ergometer) intervention study on 45 stage II breast cancer patients undergoing chemotherapy. Functional aerobic capacity ($\dot{V}O_2$max) improved 40%, maximum test time increased (subjects could work longer), and maximum workload improved (work intensity was higher) in the exercise group in comparison to patients who completed stretching and flexibility exercises or patients who remained sedentary.

Dimeo et al. (1996) found that 20 cancer patients, approximately 30 days after undergoing bone marrow transplantation, improved on maximal physical performance and maximum walking distance with significantly lower heart rates following a six-week treadmill walking program. The improvement in functional capacity was more than sufficient for carrying out daily activities. The investigators concluded that patients undergoing bone marrow transplantation experienced less fatigue and enhanced physical performance because of the exercise intervention. In 1997, Dimeo, Fetscher, Lange, Mertelsmann, and Keul conducted another investigation on bone marrow transplant patients undergoing high-dose chemotherapy. The exercise group performed interval (work-rest) exercise for 30 min daily on a supine bicycle ergometer during hospitalization. The decrement in physical performance was 27% greater in the control nonexercise group compared to the exercise group. Duration of *neutropenia* (abnormally small number of neutrophil cells in the blood) and *thrombopenia* (small number of blood platelets) was reduced in the exercise group. The investigators concluded that aerobic exercise could be safely implemented after chemotherapy with the result of improving physical performance. In another study, Mock et al. (1997) found that women exercising during radiation for breast cancer improved significantly on fuctional capacity compared to the nonexercise cancer controls. The authors concluded that a self-paced, home-based walking exercise intervention can improve physical functioning in cancer patients.

Muscular Strength

Some research has shown that exercise improves muscular strength in cancer patients. Buettner and Gavron (1981) found that patients with a history of cancer who completed an eight-week training program had significant improvements in grip strength compared with a sedentary group of cancer patients. According to Hicks (1990), cancer patients may have increased metabolic demand with concomitant impaired protein-sparing mechanisms. Inactivity, reduced caloric intake, and treatment with chemotherapy drugs can lead to loss of muscle tissue and diminished muscular endurance. Therefore, interventions such as exercise that increase muscle mass and spare protein could be highly beneficial. Muscle tension enhances protein synthesis, suggesting that exercise should be positive for cancer patients (Vandenburgh, 1987).

Butterfield and Calloway (1984) found that moderate exercise increases dietary protein utilization. Although moderate exercise appears to be beneficial, high-intensity exercise increases protein needs and therefore might not benefit cancer patients. Deuster, Morrison, and Ahrens (1986) reported that tumor growth and muscle mass loss were decreased in rats following seven weeks of endurance exercise.

In 1999, Durak, Lilly, and Hackworth integrated aerobic exercise and weight training twice per week for a 20-week period in 25 prostate cancer and leukemia patients (84% had completed treatment). The leukemia group significantly improved in strength on the bench press, hip extension, and leg press machines. The leukemia group also reported more positive results on the Modified Rotterdam Survey, in which patients were asked, for example, whether they felt stronger and whether they had more endurance.

"One of my favorite things to do is garden, and at one point I was not strong enough to garden. It was very discouraging that I had to watch other people do the planting and gardening, but after the sessions, I now have enough strength to do it myself."

Fatigue

Exercise also affects the fatigue so widespread among cancer patients (figure 4.1). Dimeo et al. (1996) found that 20 bone marrow transplantation patients experienced less fatigue and showed enhanced physical performance as a result of the exercise intervention. Dimeo, Rumberger, and Keul (1998) administered an exercise program (walking daily on a treadmill with an intensity corresponding to a lactate concentration of 3 mmol/L) to determine if cancer treatment-related fatigue could be reduced in cancer patients. Patients experienced reduced fatigue and could perform daily activities more efficiently following the exercise intervention. The investigators concluded that aerobic exercise with defined intensity, duration, and frequency could be prescribed for cancer patients to alleviate cancer- or cancer treatment-related fatigue.

Mock et al. (1997) also found that breast cancer patients had less severe symptoms of fatigue (as well as less anxiety and less difficulty sleeping) following exercise. The authors concluded that a self-paced, home-based walking exercise intervention

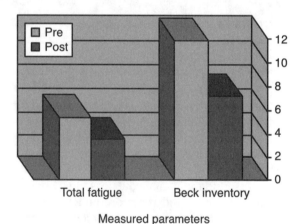

Measured parameters

Figure 4.1 Physiological and psychological variables pre- to post-intervention. Studies at the Rocky Mountain Cancer Rehabilitation Institute have shown that exercise, properly administered, can reduce fatigue and depression significantly.

helped manage treatment symptoms. In another investigation, Schwartz (2000) studied the effects of exercise on the fatigue patterns of 27 cancer patients diagnosed with stage I to stage IV breast cancer, using the 12-min walk pre- and postexercise intervention to determine functional ability. Exercise intensity was determined using the Caltrac accelerometer, which measures the energy expended in calories. The exercise intervention included self-reported exercise and self-reported daily fatigue.

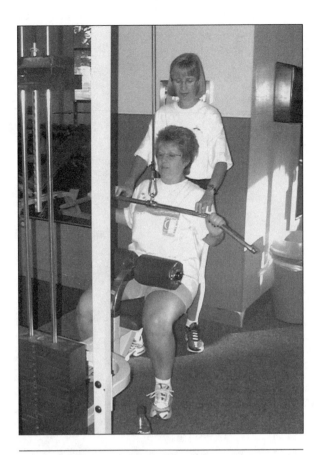

"Before cancer [exercise] rehabilitation, the only thing I could do each day was make my bed."

Schwartz found that fatigue increased sharply in the first 24 to 48 hr after chemotherapy, and some of the patients had fatigue lasting beyond five days. Whereas other investigations had indicated that fatigue was related to anemia and neutropenia, Schwartz found that fatigue did not follow hematological alterations and therefore concluded that fatigue might be caused by other mechanisms in addition to hematological alterations. Women who exercised experienced less severe fatigue and demonstrated decreased fatigue over the three courses of chemotherapy. In 2001, Schwartz, Mori, Gao, Nail, and King again investigated daily fatigue in breast cancer patients receiving chemotherapy, using self-reported measures, and found that exercise significantly reduced fatigue. As exercise duration increased, the intensity of fatigue decreased. The researchers concluded that low- to moderate-intensity exercise is effective in reducing cancer treatment-related fatigue and maintaining functional capacity. Segal et al. (2001) obtained similar results, indicating that physical exercise blunted some of the negative side effects of cancer treatment in 99 breast cancer patients.

Other Symptoms

Some research has addressed the effects of exercise on other common symptoms or concerns related to cancer or cancer treatment. It appears that besides helping with nausea and lymphedema, exercise can improve patients' ability to carry out activities of daily living, provides psychosocial benefits, and enhances quality of life.

• **Nausea.** In women with stage II breast cancer, Winningham and MacVicar (1985) found that nausea resulting from chemotherapy lessened during an exercise program (10 weeks, three times weekly, on a cycle ergometer).

• **Lymphedema.** In a study by Kalda (2000), upper body exercise did not change arm volume or arm circumference in women with lymphedema, but the women did experience an improvement in quality of life. McTiernan, Gralow, and Talbott (2000) noted that patients usually perform exercise for lymphedema while wearing the compression garments; thus the muscles are forced to contract against the wraps, which causes an outflow of lymph fluid from the arm.

• **Ability to perform customary or necessary daily activities.** In the two studies that have dealt with the effects of exercise on patients' ability to perform daily activities, the effects were positive. As noted earlier, Dimeo et al. (1996) found that exercise improved functional capacity in bone marrow transplantation patients to a level that was more than sufficient for carrying out daily activities. Later, Dimeo et al. (1998) had patients complete an exercise program (walking daily on a treadmill) and found that they could perform daily activities more efficiently.

• **Psychosocial aspects.** Exercise also affects other symptoms in cancer patients (chapter 2). Mock et al. (1994), investigating the effects of exercise in breast cancer patients undergoing chemotherapy, reported that the exercise group had higher levels of psychosocial adjustment (body image, self-concept) and lower levels of distressful symptoms (fatigue, sleeping difficulties, nausea, vomiting, depression, and anxiety). The exercise involved a moderate walking program, four to five times per week, 10 to 45 min in duration. The activity included brisk walking followed by 5 min of slow walking at the individual's own pace and desired distance. Additionally, Dimeo, Stieglitz, Novelli-Fischer, Fetscher, and Keul (1999) reported lower psychological distress (obsessive-compulsive traits, fear, interpersonal sensitivity, and phobic anxiety) in cancer patients after bone marrow transplantation compared with the nonexercise cancer patients.

• **Quality of life.** Courneya and colleagues (Courneya, Mackey, and Jones, 2000; Courneya and Mackey, 2001) synthesized much of the research in the area of cancer and exercise rehabilitation. Their conclusion, similar to that in a review of literature by the current authors, was that physical exercise had significant positive effects on the quality of life of cancer patients. The investigations reviewed included studies using self-reported exercise programs and supervised exercise interventions, studies of a variety of types of cancer (especially breast cancer), and studies of the effects of aerobic exercise and muscular strength programs. The exercise interventions ranged from 2 weeks to 20 weeks, with most lasting less than 12 weeks. The quality-of-life improvements included enhanced functional capacity, increased muscular strength, reduced fatigue, improved self-concept, enhanced vigor, and reduced anxiety and depression. Given the significance of these studies, it appears that exercise intervention is beneficial for cancer patients and should therefore be used in conjunction with treatment for cancer.

DEVELOPING EXERCISE GUIDELINES FOR CANCER PATIENTS

Winningham and colleagues (Winningham, MacVicar, and Burke, 1986; Winningham, 1991) stressed the importance of establishing guidelines that would address the needs of cancer patients who were active before their cancer diagnosis, as well as of those who had been previously sedentary. These authors stated concerns and precautions for exercise in cancer patients but did not give specific guidelines for exercise interventions (mode, intensity, etc.). General concerns that they addressed included precautions with exercise given the pathological effects that occur with cancer treatment such as fever, anorexia, weight loss, fatigue, polycythemia, anemia, leukocytosis, leukopenia, thrombocytosis, thrombocytopenia, blood coagulation abnormalities, and metabolic disorders (lactic acidosis). Winningham and colleagues emphasized that patients respond to different treatment modalities very differently and

will therefore respond to exercise training differently from each other. Cancer treatment might impair the training effects that one normally sees with an exercise intervention. The authors stated that pre-exercise screening should include screening for cardiovascular disease and orthopedic problems; determination of fitness level; consideration of type, site, and stage of cancer; screening for side effects of the cancer therapy; determination of psychological status; and careful consideration of exercise in conjunction with the timing of patients' tests and treatments.

Exercise can affect the body's physiological systems either positively or negatively, depending on its nature and extent. For example, heavy prolonged exertion is associated with hormonal and biochemical changes that potentially have a detrimental effect on immune function (Gleeson and Bishop, 2000; Nieman, Johanssen, Lee, and Arabatzis, 1990). In contrast, moderate exercise has been shown to contribute positively to the body's physiological systems, especially the immune system. Thus, you must understand how to alter your prescriptions to match each cancer patient's current condition, needs, and realistic potential to achieve the benefits exercise can provide. Guidelines for exercise intervention with this population are needed to avoid exacerbating cytotoxicities, which lead to fatigue and other symptoms, and to instead help patients mitigate or overcome the negative effects of their disease and its treatment.

As seen earlier in this chapter, research on the effects of exercise on cancer treatment-related symptoms has begun. However, most investigations have not used consistent screening, assessment, and exercise interventions designed specifically for each patient. Patients in these studies had varying types of cancers (i.e., colorectal, breast, prostate, leukemia, etc.) with different treatments (i.e., surgery, chemotherapy, bone marrow transplantation, radiation, combination therapy, etc.); yet the researchers had all patients doing the same exercise interventions with minimal pre- and post-assessment procedures. A model of an effective cancer rehabilitation program has not been specified in the current research literature.

A recent series of studies by the present authors and colleagues (Dennehy, Carter, and Schneider, 2000a, 2000b; Dennehy, Schneider, Carter, Bentz, Stephens, and Quick, 2000; Dennehy, Bentz, Stephens, Carter, and Schneider, 2000; Dennehy, Bentz, Fox, Carter, and Schneider, 1998; Schneider et al., 2000; Schneider, Dennehy, Roozeboom, and

Carter, 2002; Schneider, Bentz, and Carter, 2002a, 2002b; Stephens et al., 2000) investigated the effects of individualized prescriptive exercise intervention for cancer patients with varying types of cancer and cancer treatments. The parameters measured included physiological fatigue indexes (cardiovascular endurance, muscular strength, muscular endurance, pulmonary function, range of motion, flexibility, body composition) and depression and perception of fatigue (table 4.1). The studies involved a physical examination by a physician, pre-exercise assessment, an individualized exercise prescription, a six-month controlled exercise intervention, and a postexercise reassessment. The design of the exercise interventions in these studies was specific to each cancer patient, taking into account the type of cancer, stage of cancer, and type of cancer treatment.

The results in these investigations showed significant improvements in the patients' physiological and psychological well-being (figure 4.2). However, the most significant finding by these investigators was the critical importance of individualizing each patient's exercise according to that person's physiological and psychological status. For example, in the case of one of our patients, the physical examination identified a tumor on the inferior vena cava. The physician reported that the patient had the potential to be harmed by strenuous activity; therefore the exercise assessment and prescription had to be modified substantially in order to provide help and not do harm. In another situation, the physical examination of one of our patients identified severe cardiovascular disease. Physical exercise could have exacerbated potential cardiac events. Our patients with periph-

Table 4.1   Data from the Rocky Mountain Cancer Rehabilitation Institute

Variable	Exercise group			Control group		
	Pre-	Post-	% change	Pre-	Post-	% change
$p\dot{V}O_2$ (ml · kg^{-1} · min^{-1})	20.4	24.40	19.6%	21.3	24.47	14.8%
Time on treadmill (minutes)	5.28	7.17	35.8%	5.18	6.3	21.6%
Resting heart rate (bpm)	83.2	78	−6.3%	81.0	80.6	−1.2%
AbS (number of sit-ups)	18.0	29.4	63.3%	22.0	24.8	12.7%
Total fatigue (subjective scores)	4.97	2.89	−41.9%	4.67	3.36	−28.0%
Beck Depression	10.09	5.56	−44.9%	13.47	9.38	−30.3%

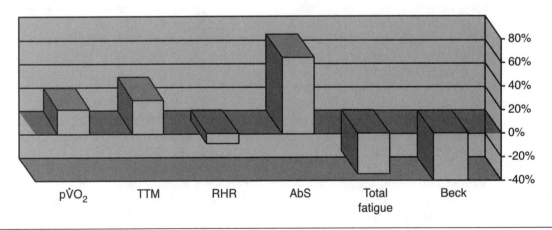

Figure 4.2   Percent change in physiological and psychological variables in RMCRI patients, pre- to post-intervention. $p\dot{V}O_2$ = predicted $\dot{V}O_2$max; TTM = time on treadmill; HRrest = resting heart rate; AbS = abdominal strength.

eral neuropathy and vestibular abnormalities have a high risk of losing their balance, so the exercise intervention must be individualized to prevent injury.

SOME GENERAL RESEARCH-BASED GUIDELINES

Because the use of exercise as a therapeutic and rehabilitative tool is in its infancy, only four studies have offered general guidelines for cancer patients as of this writing.

1. **Hicks (1990)** described exercise goals that should be implemented in early and late stages of cancer:

   - **Early stages.** Goals for patients in the early stages of cancer should be preventive and restorative. Because the patient might have lost strength or flexibility as a result of treatment, restoring strength and range of motion becomes the goal. Hicks recommended endurance exercise (e.g., exercise bike) following careful screening.

   - **Late stages.** Patients with late-stage cancer have more severe deficits because of the cancer itself and the cancer treatment side effects. Hicks suggested tailoring a safe program for patients to help increase strength, range of motion, and endurance. Screening, according to Hicks, should include determination of the fitness level of the patient, identification of coexisting medical problems, and careful analysis of the cancer condition.

2. **Winningham (1994)** suggested the following components of the exercise prescription for cancer patients:

   - The cancer patient's medical and functional status should be screened and analyzed. The screening should include the type of cancer, disease progression, and treatment effects.

   - Intensity and duration are standard for cancer patients who have a performance status of 0 to 1 (0 = normal activity, no restrictions; 1 = symptomatic but ambulatory, able to engage in self-care) on the Zubrod functional independence scale.

   - Patients with some functional impairment (2 on Zubrod scale = ambulatory >50% of time; needs occasional assistance) may benefit from an exercise program that incorporates brief-duration exercise such as 5- to 10-min walk-rest intervals. Exercise may be more intense if monitoring and supervision are available.

   - The exercise mode should be based on safety, convenience, and the patient's preference. For example, high-impact activities such as rope jumping and jogging would not be appropriate for patients with bone metastases because of the possibility of stress fractures, or for patients with low platelet counts because of the possibility of bleeding into joint cavities. Walking and cycle ergometers are the preferred modes of exercise with high intensity to low intensity interval work.

   - Patients who are more debilitated should use less intensity and engage in more frequent workout sessions per week.

3. **Mock (1996)** proposed the following approach to exercise prescription for breast cancer patients:

   - Prescreening is essential before the development of an exercise program. Screening should include fitness level, age, stage of cancer, treatment, and other general health considerations.

   - Each breast cancer patient's exercise program should be individualized because people may be at varying fitness levels prior to cancer diagnosis.

   - The patient should exercise four to six days per week, should not skip more than one day in a row, and should take one rest day per week.

   - Exercise should be at an intensity of 60% to 70% of target heart rate (100-120 bpm) at a self-paced brisk walk for 10 to 30 min.

   - Cancer patients should progress to a longer duration as exercise tolerance increases.

   - A patient on chemotherapy that might result in cardiotoxicity requires supervision during exercise.

4. **Courneya and Mackey (2001)**, after reviewing the literature, suggested these exercise prescription guidelines:

   - Exercises should involve large-muscle groups—examples are walking and cycling.

- Frequency should be three to five times per week, but daily exercise could be more beneficial.

- Intensity should be moderate, corresponding to 50% to 75% of heart rate reserve.

- Duration should be at least 20 to 30 min of continuous exercise but may have to occur in shorter bouts for poorly conditioned individuals.

- Progression should include increases in frequency and duration first, followed by increases in intensity as the patient becomes more fit. The progression should be gradual, especially for patients with severe side effects of treatment.

RMCRI GUIDELINES FOR PRESCRIPTIVE EXERCISE IN CANCER PATIENTS

The Rocky Mountain Cancer Rehabilitation Institute has been investigating the effects of exercise intervention on cancer patients since 1996. The Institute's goal has been to develop a model for cancer rehabilitation patients that includes appropriate assessment, prescription, and program intervention. Over the past six years, the program designed at RMCRI has proven effective in increasing the physiological and psychological well-being of cancer patients (Schneider et al., 2002, 2002a, 2002b). Upon reviewing the current literature on guidelines just described, we found that the guidelines may work for a couple of cancer patients but certainly not for everyone. For example, some of our patients many years out of treatment could do 20 min of continuous exercise, but many patients in the process of undergoing cancer treatment could not do even 10 min of continuous activity. We also found that the exercise prescription could not be linear (gradual progression). Each exercise session might vary depending on the health status of the patient. A patient might be able to do 15 min of activity on some days but on other days, after having just had a chemotherapy treatment, be able to do only very low intensity activity for 5 min. Therefore, we saw the need for individualizing each exercise prescription for each patient.

Thus far our experience has demonstrated that the essential components of cancer rehabilitation include a physician referral, a physical examination, appropriate screening, physical and psychological assessments, individualized exercise prescriptions, a six-month individualized exercise intervention program, reassessment, and community-based ancillary services. Physician referral took on importance in our program because we wanted the exercise assessment and intervention to become part of the patient's cancer treatment. The only way to incorporate exercise into the cancer treatment is to work in collaboration with the physicians. Experience also showed us the importance of a physical examination. Cancer patients appear healthy because you usually cannot detect internal toxicities that are present. An outstanding example, referred to earlier, was the patient whose physical examination revealed a tumor on the vena cava. The physician informed us that the patient could "bleed out" with a heavy bout of exercise. Patient screening is also extremely important in cancer rehabilitation: because most cancer patients have multiple physical conditions, you need to screen for multiple physical problems. Finally, many investigations in the area of cancer rehabilitation have not used pre- and post-assessments. In order to determine the effectiveness of your intervention, you must have quantifiable assessment parameters. The assessments must use measurements that are appropriate (that assess cancer toxicities) and that are reliable and valid.

From the beginning we learned that patients could not progress like healthy individuals in an exercise program. We also found that we could not write the same exercise prescription for all patients. Therefore, it was necessary to develop individualized exercise prescriptions according to the patient's health and treatment status. With the individualized exercise prescriptions came the need to develop individualized exercise program interventions. Each patient has his own cancer-related physical and psychological problems. You cannot develop an exercise program implementing fine motor skills if the patient has balance problems—instead you must work carefully with the patient on restoring balance. We chose six months as the length of the program so that there would be plenty of time for the patient to show improvement and get accustomed to the program and thus continue exercise beyond our program. Reassessment or postexercise intervention assessment is necessary to determine the effectiveness of a program. Ancillary services became necessary because although we wanted to help the patients in all areas, our expertise is in exercise. Therefore we developed a network of ancillary services (i.e., massage, psychotherapy, etc.) so we could refer patients who needed services that we could not provide.

The key to cancer rehabilitation is not only to have a basic program for assessment, exercise prescription, and intervention, but also to individualize the components to fit each patient. Chapters 5 through 7 provide details of the guidelines currently used at RMCRI for assessment, prescription, and intervention and include discussion of why we feel these guidelines are appropriate.

The following is a brief explanation of each essential component of a cancer rehabilitation program. Chapters 5 through 7 describe the components in detail.

• **Pre-assessment procedures.** Physician referral is essential to the cancer rehabilitation program. In many instances, as already noted, cancer patients appear healthy because many of the symptoms associated with cancer and cancer treatments are not obvious; yet these symptoms can be serious, even life threatening. In addition to having a physician referral, it is essential that you meticulously gather information from the patient prior to the assessment. From this information, you can determine appropriate assessment techniques and protocols.

• **Pre-assessment modifications.** Based on the information obtained by the referring physician, the information obtained by the physician during the physical exam, and the information supplied by the patient, the appropriate assessment tests and procedures are selected and completed. Not all assessment procedures are pertinent for all cancer patients.

• **Assessment procedures.** Certain assessments are essential to enable you to monitor the factors that you need to watch in order to mitigate the physiological and psychological side effects of cancer treatment (Schneider et al., 2002, 2002a, 2002b). These assessments should therefore be considered the standard model (see "Standard Assessments for Cancer Patients in Exercise Therapy"). Alterations to this standard should be based on specific limitations of the patient.

• **Post-assessment procedures.** Following the standard assessments for cancer patients in exercise therapy, the patient receives instructions regarding the development of the exercise prescription. The dismissal of the patient should occur only after the heart rate and blood pressure have reached near-resting values.

Standard Assessments for Cancer Patients in Exercise Therapy

- Informed consent
- Medical history and cancer history
- Fatigue, quality-of-life, and psychological indexes
- Cardiovascular disease risk factors
- Physician physical exam
- Physiological assessments in order of administration
    1. Resting and exercise heart rate
    2. Resting and exercise blood pressure
    3. Pulse oximetry (oxygen saturation)
    4. Body composition (skinfolds)
    5. Pulmonary function
    6. Circumference measurements
    7. Cardiorespiratory endurance: physician present
    8. Flexibility
    9. Range of motion
    10. Muscular endurance
    11. Muscular strength
- Three-day diet analysis

• **Individualized exercise prescription.** The cancer exercise specialist develops the individualized exercise prescription after careful analysis of the patient's type of cancer, cancer treatment, and length of time out from treatment. The prescription is divided into three broad categories:

- Fitness assessment results, medical exam, and cancer history
- Exercise recommendations and precautions
- Six-month goals

• **Six-month exercise intervention.** Trained cancer exercise specialists should be responsible for carefully designing the six-month exercise intervention program from the exercise prescription. A six-month intervention allows the cancer patient sufficient time for physiological improvements. The three day per week program should accomplish the intended goals and improve the patient's functional capacity. Modifications of the exercise prescription are made on the basis of changes in treatment regimens or drug therapy, or as a result of any new developments in health status. The rate of progression is in turn based on these modifications in the patient's exercise prescription. The exercise intervention will differ significantly between the patient's treatment period and the posttreatment period. The variations in programming between patients are addressed in chapter 6.

• **Reassessment.** Following the six-month exercise intervention program, the cancer patient again completes all the assessment procedures that were used before the exercise intervention. The differences between the pre-assessment values and the reassessment values help determine the effectiveness of the exercise intervention. The reassessment also helps determine modifications in the exercise prescription for the next six-month exercise intervention.

• **Ancillary services.** Services within the community should be available to meet the special needs of patients. Services may include psychological counseling, physical therapy, massage therapy, and occupational therapy, as well as provision of information on other applicable services throughout the community.

SUMMARY

A growing body of research is showing that exercise interventions can help cancer patients during treatment and recovery. Studies on exercise benefits have documented improvements in aerobic capacity and muscular strength in cancer patients during and after various types of therapies. Exercise also helps with the fatigue that many cancer patients experience and with other common cancer treatment-related symptoms and concerns. Besides alleviating nausea and lymphedema, it appears that exercise improves patients' ability to carry out activities of daily living, affords psychosocial benefits such as improved body image and self-concept, and enhances quality of life.

Guidelines for exercise intervention for cancer patients are needed, and the cancer exercise specialist must understand how to tailor prescriptions to specifically match each patient's current physiological and psychological status, needs, and goals. The literature contains four studies (Courneya and Mackey, 2001; Hicks, 1990; Mock, 1996; Winningham, 1994) that present general re-

search-based guidelines for cancer patients, suggesting appropriate goals and exercise prescription components. The present authors' ongoing investigation of the effects of exercise intervention on cancer patients has shown that it is essential to individualize each exercise prescription for each patient. This research also suggests that the essential components of rehabilitation include a physician referral, a physical examination, appropriate screening, physical and psychological assessments, individualized exercise prescriptions, a six-month individualized exercise intervention program, reassessment, and community-based ancillary services.

STUDY QUESTIONS

1. What does the current literature suggest concerning the development of aerobic capacity and muscular strength in cancer patients?

2. What are the psychological benefits of exercise for cancer patients?

3. Why is it important to develop exercise guidelines for cancer patients?

4. What are some general exercise guidelines for cancer patients?

5. What are the essential components for a cancer rehabilitation program?

REFERENCES

Buettner, L.L., and S.J. Gavron. 1981. *Personality changes and physiological effects of a personalized fitness enrichment program for cancer patients.* Paper presented at the Third International Symposium on Adapted Physical Activity, New Orleans.

Butterfield, G.E., and D.H. Calloway. 1984. Physical activity improves protein utilization in young men. *Br J Nutr* 51:171–187.

Courneya, K.S., and J.R. Mackey. 2001. Exercise during and after cancer treatment: Benefits, guidelines, and precautions. *Int Sport Med J* 1(5):1–8.

Courneya, K.S., J.R. Mackey, and L.W. Jones. 2000. Coping with cancer. *Physician Sportsmed* 28(5):49–73.

Dennehy, C.A., A. Bentz, K. Fox, S.D. Carter, and C.M. Schneider. 1998. Breast cancer risk profile of sedentary, active, and highly trained women. *Med Sci Sports Exerc* 30(5):S63.

Dennehy, C.A., A. Bentz, K. Stephens, S.D. Carter, and C.M. Schneider. 2000. The effect of cancer treatment on fatigue indices. *Med Sci Sports Exerc* 32(5):S234.

Dennehy, C.A., S.D. Carter, and C.M. Schneider. 2000a. Physiological manifestations of prescriptive exercise on cancer treatment-related fatigue. *Physiologist* 43(4):357.

Dennehy, C.A., S.D. Carter, and C.M. Schneider. 2000b. Prescriptive exercise intervention for cancer treatment-related fatigue. *Physical Activity and Cancer: Cooper Aerobics Institute,* No. 4, p. 22.

Dennehy, C.A., C.M. Schneider, S.D. Carter, A. Bentz, K. Stephens, and K. Quick. 2000. Exercise intervention for cancer-related fatigue. *Res Q Exerc Sport* 71(1, suppl):A-27.

Deuster, P.A., S.D. Morrison, and R.A. Ahrens. 1986. Endurance exercise modifies cachexia of tumor growth in rats. *Med Sci Sports Exerc* 3:385–392.

Dimeo, F., H. Bertz, J. Finde, S. Fetscher, R. Mertelsmann, and J. Keul. 1996. An aerobic exercise program for patients with haematological malignancies after bone marrow transplantation. *Bone Mar Trans* 18:1157–1160.

Dimeo, F., S. Fetscher, W. Lange, R. Mertelsmann, and J. Keul. 1997. Effects of aerobic exercise on the physical performance and incidence of treatment-related complications after high-dose chemotherapy. *Blood* 90:3390–3394.

Dimeo, F., B.G. Rumberger, and J. Keul. 1998. Aerobic exercise as therapy for cancer fatigue. *Med Sci Sports Exerc* 30(4):475–478.

Dimeo, F., R. Stieglitz, U. Novelli-Fischer, S. Fetscher, and J. Keul. 1999. Effects of activity on fatigue and psychologic status of cancer patients during chemotherapy. *Cancer* 85:2273–2277.

Durak, E.P., P.C. Lilly, and J.L. Hackworth. 1999. Physical and psychosocial responses to exercise in cancer patients: A two year follow-up survey with prostrate (sic), leukemia, and general carcinoma. *JEPonline* 2(1). Available at http://www.css.edu/users/tboone2/asep/jan12b.htm. Accessed May, 2000.

Gleeson, M., and N.C. Bishop. 2000. Elite athlete immunology: Importance of nutrition. *Int J Sports Med* 21(suppl 1):S44–S50.

Hicks, J.E. 1990. Exercise for cancer patients. In *Therapeutic exercise,* ed. J.V. Basmajian and S.L. Wolf. Baltimore: Williams & Wilkins.

Kalda, A.L. 2000. The effect of upper body exercise on secondary lymphedema following breast cancer treatment. Microform Publications, University of Oregon, Eugene, OR.

MacVicar, M.G., M.L. Winningham, and J.L. Nickel. 1989. Effects of aerobic interval training on cancer patients' functional capacity. *Nurs Res* 38(6):348–351.

McTiernan, A., J. Gralow, and L. Talbott. 2000. *Breast fitness.* New York: St. Martin's Press.

Mock, V. 1996. The benefits of exercise in women with breast cancer. In *Contemporary issues in breast cancer,* ed. K.H. Dow. Sudbury, MA: Jones & Bartlett.

Mock, V., M.B. Burke, P. Sheehan, E.M. Creaton, M.L. Winningham, S. McKenney-Tedder, L.P. Schwager, and M. Liebman. 1994. A nursing rehabilitation program for women with breast cancer receiving adjuvant chemotherapy. *Oncol Nurs Forum* 21:899–907.

Mock, V., K.H. Dow, C.J. Meares, P.M. Grimm, J.A. Dienemann, M.E. Haisfield-Wolfe, W. Quitasol, S. Mitchell, A. Chakravorthy, and I. Gage. 1997. Effects of exercise on fatigue, physical functioning, and emotional distress during radiation therapy for breast cancer. *Oncol Nurs Forum* 24(6):991–1000.

Nieman, D.C., L.M. Johanssen, J.W. Lee, and K. Arabatzis. 1990. Infectious episodes in runners before and after the Los Angeles Marathon. *J Sports Med Phys Fit* 30:316–328.

Schneider, C.M., A. Bentz, and S.D. Carter. 2002a. *The influence of prescriptive exercise rehabilitation on fatigue indices.* Manuscript in preparation.

Schneider, C.M., A. Bentz, and S.D. Carter. 2002b. *Prescriptive exercise rehabilitation adaptations in cancer patients.* Manuscript in preparation.

Schneider, C.M., C.A. Dennehy, M. Roozeboom, and S.D. Carter. 2002. A model program: Exercise intervention for cancer rehabilitation. *J Integr Cancer Ther* 1(1):76–82.

Schneider, C.M., K. Stephens, A. Bentz, K. Quick, S.D. Carter, and C.A. Dennehy. 2000. Prescriptive exercise rehabilitation adaptations in cancer patients. *Med Sci Sports Exerc* 32(5):S234.

Schwartz, A.L. 2000. Daily fatigue patterns and effect of exercise in women with breast cancer. *Cancer Pract* 8(1):16–24.

Schwartz, A.L., M. Mori, R. Gao, L.M. Nail, and M.E. King. 2001. Exercise reduces daily fatigue in women with breast cancer receiving chemotherapy. *Med Sci Sports Exerc* 33(5):718–723.

Segal, R., W. Evans, D. Johnson, J. Smith, S. Colletta, J. Gayton, S. Woodard, G. Walls, and R. Reid. 2001. Structured exercise improves physical functioning in women with stages I and II breast cancer: Results of a randomized controlled trial. *J Clin Oncol* 19(3):657–665.

Stephens, K., C.M. Schneider, A. Bentz, M. Lapp, K. Quick, S.D. Carter, and C.A. Dennehy. 2000. The influence of time from cancer treatment on selected fatigue indicators. *Med Sci Sports Exerc* 32(5):S233.

Vandenburgh, H.H. 1987. Motion to mass: How does tension stimulate muscle growth? *Med Sci Sports Exerc* 19:142–149.

Winningham, M.L. 1991. Walking program for people with cancer: Getting started. *Cancer Nurs* 14:270–276.

Winningham, M. 1992. The role of exercise in cancer therapy. In *Exercise and disease,* ed. M. Eisinger and R.W. Watson, 63–70. Boca Raton, FL: CRC Press.

Winningham, M.L. 1994. Exercise and cancer. In *Exercise for prevention and treatment of illness,* ed. D.L. Elliot and L. Goldberg. Philadelphia: Davis.

Winningham, M.L., and M.G. MacVicar. 1985. Response of cancer patients on chemotherapy to a supervised exercise program [Abstract]. *Med Sci Sports Exerc* 17:292.

Winningham, M.L., M.G. MacVicar, and C. Burke. 1986. Exercise for cancer patients: Guidelines and precautions. *Physician Sportsmed* 14:125–134.

5

Health and Fitness Assessment for Cancer Patients

An initial assessment is necessary for determining the cancer patient's strengths and weaknesses. This assessment also provides baseline data for the development of the individualized exercise program. Moreover, regular assessment of the patient's physiological and psychological parameters allows you to monitor improvements, make changes in the exercise prescription, recommend ancillary services, and meet specific needs of the patient. Comparisons between the initial assessment and subsequent reassessments provide information on the effectiveness of the exercise intervention program. No established assessment procedures specific to cancer patients exist at present, although the Rocky Mountain Cancer Rehabilitation Institute (RMCRI) is currently developing assessment techniques that will have been validated on cancer patients. However, the need for professionally based cancer rehabilitation is immediate, and emerging programs must rely on available protocols until population-specific equations and techniques have been developed.

This chapter begins by establishing the foundational principles of RMCRI's assessment program and moves on to present the details of the three phases of that program: the pre-assessment, assessment, and post-assessment stages. Much of the research base for the program appears in publications emanating from our own work (Dennehy, Carter, and Schneider 2000a, 2000b; Dennehy, Schneider, Carter, Bentz, Stephens, and Quick, 2000; Dennehy, Bentz, Stephens, Carter, and Schneider, 2000; Dennehy, Bentz, Fox, Carter, and Schneider, 1998; Schneider et al., 2000; Schneider, Dennehy, Roozeboom, and Carter, 2002; Schneider, Bentz, and Carter, 2002a, 2002b; Stephens et al., 2000).

PRINCIPLES OF EFFECTIVE ASSESSMENT

The Rocky Mountain Cancer Rehabilitation Institute adheres to six essential principles of effective assessment for cancer patients:

1. **Each parameter measured during an assessment should be relevant to the improvement of the cancer patient's tolerance for treatment, recuperation from treatment, and overall health status.** This means that the assessment must be as specific as possible to the patient's present health status.

2. **All procedures used should be valid and reliable.** A test is valid when it actually measures

what it claims to measure. The test is reliable when the results are consistent and reproducible.

3. **Administration of each assessment must be consistently and rigidly controlled.** This necessitates reliability testing for the staff, standardization of instructions to all patients, exact preparation procedures, same order of test items, same recovery time between tests, same calibrated equipment, controlled environment, and same pretest instructions to the patients.

4. **The patient's rights should be protected.** The cancer exercise specialists should carefully explain the nature of the tests, as well as any risk involved during the tests, and assure the patient that all results are appropriately charted and remain confidential.

5. **Testing should be repeated at regular intervals, such as every six months.** The main purpose of the testing is to develop an effective individualized exercise intervention. Therefore, assessments must be repeated at specified intervals during the rehabilitation process to allow monitoring of progress as well as review appropriate modifications for the intervention.

6. **The assessments should be reviewed and explained to the patient.** The results are used to develop the individualized exercise prescription. The results should be understood by both the cancer exercise specialist and the cancer patient.

PRE-ASSESSMENT PROCEDURES

The pre-assessment procedures involve preparing for the administration of each protocol and preparing to receive the patient. These steps help to ensure that information gathered prior to and during the assessment procedure is thorough and accurate. Pre-assessment information is valuable to the physician conducting the physical examination, allowing her to make recommendations about assessment procedures. These recommendations may involve modifying protocols or omitting certain activities that would be contraindicated for a particular patient. Additionally, the cancer exercise specialist uses this information when developing the exercise prescription.

Reliability Testing for Cancer Exercise Specialists

Cancer exercise specialists are trained to work specifically with cancer patients during and following treatment. To ensure accuracy and consistency during assessments, each specialist involved in the assessment of patients should undergo reliability testing. *Reliability* testing ensures that the data collected on each patient before and after the exercise intervention are accurate and that consistent patterns will be used in the techniques so reliable data are obtained. During reliability testing for evaluation of accuracy, the cancer exercise specialists must repeat each test protocol (on healthy subjects) until they obtain a reliability of $r = .95$ (RMCRI criterion) or better on the protocol procedures and results before working with cancer patients.

Physician Referral

New patients entering the program should have a referral prescription from their primary care physician or oncologist. A referral system ensures that the physicians and cancer exercise specialists work together to provide quality care. Once the patient's physician refers the patient, the medical director at RMCRI and the cancer exercise specialists work together to communicate with the referring physician about problems, ancillary services, progress, and so on.

Pre-Assessment Mailings

Once the patient has been referred, information should be gathered from the patient on cancer and medical history, diet, depression, fatigue, quality of life, and lifestyle. The patient's blood profile information is obtained from the referring physician with the patient's permission. The other information is obtained from the cancer patient. It is useful to mail to the patient ahead of time those forms that require a lengthy amount of time to complete. (Appendix A contains forms developed at RMCRI that readers should feel free to photocopy. For diet, depression, fatigue, and quality of life, any of a number of standard, reliable forms may be used.) Also included in this mailing should be information regarding the procedures and what to expect during the assessment process. Ideally, the mailing should include the items on the list that follows. Request that the patient carefully read the informational pieces and thoroughly fill out all the forms before arriving at the cancer exercise rehabilitation facility.

• **Procedural information for patients.** This form provides information for patients on the procedures that will be followed during the assessment and reassessment.

• **Diet analysis.** The patient should complete a three-day diet analysis before arriving. The diary allows the patient to itemize the types and quantities of foods eaten. Dieticians analyze these data and make recommendations regarding the proper intake of food groups, vitamins, minerals, and water. In addition, caloric analysis can reveal any deficiencies that may precipitate or exacerbate cancer treatment-related symptoms.

The information the patient provides on the following forms is vital when it comes to writing an effective individualized exercise prescription:

• **Cancer and medical history forms.** These forms call for information such as cancer type; surgical interventions; postsurgery treatment; complications; current medical concerns; medica-

tions; and the names of the primary care physician, surgeon, oncologist, and radiation oncologist.

• **Depression inventory.** The patient's answers to the questions on this form are used to quantify the individual's state of depression. This information is important because depression can add to a patient's degree of fatigue. For any patient who is experiencing heightened depression, you may want to recommend professional counseling.

• **Subjective fatigue scale.** This questionnaire gathers information about the current level of fatigue and ways in which fatigue interferes with daily activities. The fatigue scale should evaluate various subjective physiological and psychological parameters.

• **Quality-of-life inventory.** This questionnaire calls for information about the patient's current quality of life. The inventory should include subjective indexes of daily living.

• **Lifestyle evaluation.** This form elicits information regarding the daily activities and lifestyle of the patient, including smoking, alcohol consumption, sleep patterns, exercise, and recreation.

The patient can take time to fill out the forms properly if they are mailed to the patient's home.

Patient Chart Preparation and Data Management

A chart is prepared for each patient prior to the first assessment. The patient chart will eventually contain information gathered by the physician during the physical examination and by the cancer exercise specialist during pre-assessment, assessment, development of the exercise prescription, and reassessment. All pertinent information regarding the patient and the rehabilitation intervention should be maintained in the chart. Any changes in program, exercise prescription, medications, or health status will also be recorded in the chart. The cancer rehabilitation physician (medical director) should retain ready access to the chart and can write recommendations for the cancer exercise specialist regarding any changes in the intervention. If patients have problems or if their health status changes, the referring physician will be contacted. The chart includes the following documents:

- **Chart contents.** This form indicates the order in which the forms appear in the file (see appendix A).
- **Charting sheet.** The charting sheet is for entering any pertinent patient information not included in the standard chart forms. This type of information includes the date the prescription was presented to the patient, changes in medication or health status, and the time needed before reassessment.
- **Assessment checklist.** The assessment checklist is for recording the completion of all assessment requirements.
- **Fatigue scale calculations.** This form is used to calculate fatigue levels from the fatigue scale filled out by the patient.
- **Informed consent.** The consent form explains the risks involved in the assessment process and the right of the patient to withdraw at any time. The patient, by signing, indicates that he has a full understanding of the risks and is willing to proceed.
- **Physical exam form.** This general physical exam form is used by the physician (medical director) during the initial assessment and the reassessment.
- **Cancer history.** This form provides information about cancer type, location, surgery, adjuvant treatment, medications, and any concerns due to cancer.
- **Medical history.** The medical history form, filled out by the patient, provides current and past medical information, family history, and information about operations and medications.
- **Cardiovascular disease risk factors.** Primary risk factors for coronary heart disease are quantified and discussed with the patient.
- **Data collection sheet.** The data collection sheet is for recording data obtained during each of the testing protocols.

ASSESSMENT PROCEDURES

Cancer patients, of course, are not healthy, and therefore cannot be assessed as if they were healthy. On the other hand, evaluating these patients for the purpose of prescribing exercise is a new enough phenomenon that normative data are not yet available. Such data are being developed at RMCRI, but that process is not complete. Therefore, this chapter supplies normative data for the general population only. You should keep in mind as you look at these data that the fitness categories are not appropriate for cancer patients—nevertheless, these norms are a place to start. The most important aspect of the assessment is that it gives you baseline values for comparison with the reassessment values. For the cancer patient, the percent improvement is the most critical element.

The medical experts at RMCRI suggest that modifications in the assessment procedures be considered if the patient has significant hematological abnormalities (e.g., symptomatic thrombocytopenia, neutropenia, and anemia), new undiagnosed medical conditions (e.g., *cardiac dysrhythmias* and chest pain), neurological abnormalities (e.g., disorientation), bone pain, electrolyte disturbances (severe nausea/vomiting), muscle weakness, or fever. Accordingly, we first discuss nonmodified assessment procedures and then outline the modifications that you should consider in patients with these conditions.

Preliminary Procedures

The assessment appointment begins with several preliminary steps that bring together various types of patient information. These steps include review of screening information, measurement of resting heart rate and blood pressure, resting *pulse oximetry,* review of blood values obtained by the patient's physician, and the physical examination by the physician associated with the cancer rehabilitation facility.

• **Patient arrival.** Patients arriving at the setting should find it a welcoming place. Before a patient arrives, the cancer exercise specialist should be familiar with the type of cancer the person has and with that individual's treatments. The cancer exercise specialists must not be surprised by any patient's appearance or deformities. Patients feel comfortable when the specialist knows their name and basic cancer information about them. We get a basic cancer history from the patient when we call to set up the appointment for the assessment. The setting should be clear and spacious and should be limited to cancer patients and those working with cancer patients. Patients who have body image problems are very uncomfortable around healthy, young persons. The exercise specialist should be dressed neatly in appropriate clothing such as sweats and tennis shoes. We have printed shirts (RMCRI and logo) for all our cancer exercise specialists, which sets a very professional and positive tone. As an exercise specialist you should always speak in a positive manner and give all your attention to the patients.

• **Screening.** Patients bring with them the forms they received in the mail (listed earlier), filled out to the best of their ability. The cancer exercise specialist and the patient carefully review the screening forms to ensure accuracy. Additionally, patients complete an informed consent and cardiovascular disease risk factor profile. Again, you need to pay careful attention to the cancer history. Accuracy with regard to type of cancer, stage of cancer, treatments, medications, and complications is essential. Be sure that complete information is available to the cancer rehabilitation facility's physician for the physical examination. If patients cannot recall all the information needed, we ask permission to call their personal physician.

• **Resting heart rate.** Pretest (resting) heart rate (bpm) is measured by palpation or heart rate monitor. Auscultation or ECG (electrocardiogram) readings may be used. Be sure that upon arrival the patient sits for 5 to 10 min so that she becomes accustomed to the environment and readings have time to stabilize.

• **Blood pressure (mmHg).** Again, make sure the patient has had time to get accustomed to the environment. Working with patients on the screening paperwork calms patients before the measurement of blood pressure. Take blood pressure on the right side of the body unless the patient has had surgery for breast cancer on that side, in which case take blood pressure on the unaffected side. Table 5.1 provides resting blood pressure norms.

Table 5.1   Resting Blood Pressure (BP) Norms

Classification of blood pressure for adults aged 18 years and older			
Category	**Systolic BP (mmHg)**		**Diastolic BP (mmHg)**
Optimal	<120	and	<80
Normal	120-129	and	80-84
High normal	130-139	or	85-89
Hypertension			
Stage 1	140-159	or	90-99
Stage 2	160-179	or	100-109
Stage 3	180	or	110

Ensuring Pulse Oximeter Accuracy

During assessments, use a handheld pulse oximeter. Note the following conditions that will skew the accuracy of the readings:

- Excessive motion
- Improperly attached sensor
- Pressing the sensor against any surface and squeezing or holding the sensor
- Anemia or low hemoglobin concentrations
- Sensor not at heart level

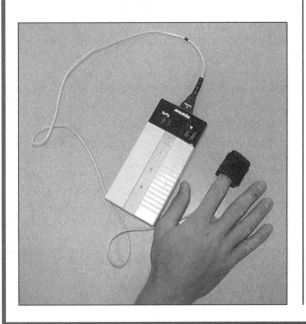

- Presence of fingernail polish on the patient (may reduce light transmission, which may affect values). In the initial instructions sent to patients, inform them to remove fingernail polish.
- Placing the blood pressure cuff on the same arm as the pulse oximeter
- Finger thickness less than 0.3 in. (0.76 cm) or greater than 1.0 in. (2.54 cm).

During cancer rehabilitation sessions, use the finger pulse oximeter. To avoid the presence of conditions that will affect the accuracy of the readings, follow these guidelines:

- Avoid excessive or rapid movement, which may affect measurements.
- While it is on the finger, do not press the monitor against any surface and do not squeeze or hold it.
- Note that fingernail polish may reduce light transmission, which may affect values.
- Make sure that the finger has adequate perfusion (do not have the finger pressing against an immovable object).
- Do not place the blood pressure cuff on the same arm as the pulse oximeter.
- Finger thickness should be between 0.3 and 1.0 in. (0.76 cm and 2.54 cm).

• **Resting pulse oximetry.** Assess the percent of oxygen saturation in the patient's blood using a handheld pulse oximeter during assessments and a finger pulse oximeter during cancer exercise rehabilitation sessions. Follow the manufacturer's instructions for each. Keep in mind the conditions that will affect the accuracy of the readouts (see "Ensuring Pulse Oximeter Accuracy").

• **Resting blood values.** At the time of assessment you should have blood values (hemoglobin, white blood cell counts, platelet counts, etc.) that will help you more fully understand the health of the patient. Blood values can be obtained from the patient's physicians with the patient's permission, or patients can bring the blood values with them when they come to the assessment.

• **Height and body weight.** Obtain height (inches; no shoes) and weight (pounds; no shoes); you should use the same height and weight instruments each time you assess the patient.

• **Physical examination.** A physician associated with the cancer exercise rehabilitation center examines the patient's heart and lung sounds, reflexes, proprioception, surgical incisions, and ports. The patient-physician consultation includes further exploration of the screening information such as specific medications, problems, and complications. The goal is to ensure that the screening information is complete and accurate. Following the examination, the physician informs the cancer exercise specialist, with the patient present, of any alterations that must be made to the remaining assessment procedures. Giving the patient plenty of information adds to the positive perception of the program. The physical examination takes place on the same day as the rest of the assessment procedures.

Physiological Assessment Parameters

A number of physiological variables can serve as indexes for fatigue, including cardiovascular endurance, muscular strength, muscular endurance, flexibility, range of motion (poor flexibility and range of motion add to energy expenditure), pulmonary function, and blood parameters. More research is needed to determine the relationship between the perception of fatigue and the quantifiable physiological variables that are possible fatigue indexes in cancer patients.

The exercise testing protocols (Heyward 1998, 2002) should be performed in the following order:

1. Body composition
2. Pulmonary function
3. Circumference measurements
4. Cardiovascular endurance
5. Flexibility, range of motion, agility
6. Muscular endurance and strength

Body Composition (Skinfolds)

We use a three-site skinfold test (triceps, suprailiac, and abdominal sites) to measure body composition. The thigh is substituted for the abdomen for women who have had TRAM-flap reconstructive surgery. Other site modifications may be required for patients with varying types of ports and prostheses. Specific regression formulas for men and women are used to determine the body fat percentage. When male specialists are performing the test on a female, a female witness must be present. When a female specialist is performing skinfold measurement on a male, a male witness should be present. Hydrostatic weighing is not used because the majority of cancer patients are susceptible to infection. The risk of infection is high if the water and underwater weighing tank are not absolutely clean. Bioelectrical impedance is not used because of the lack of accuracy, especially since cancer patients are often dehydrated. The purpose of assessing body composition is to monitor alterations in fat and lean mass that may occur as a result of treatment toxicities.

Although the three-site formula using triceps, suprailiac, and abdomen for both women and men

is ideal, TRAM-flap reconstruction requires the use of triceps, suprailiac, and thigh. For non-breast cancer patients, take all measurements on the right side of the body. For breast cancer patients, take measurements on the *nonsurgery* side of the body. The following are general guidelines for measurement:

- Place caliper 1 cm away from the thumb and finger, perpendicular to the skinfold, and halfway between the crest and the base of the fold.
- Maintain the pinch while reading the caliper.
- Wait 1 to 2 sec (no longer) before reading the caliper.
- Rotate through all of the measurement sites once.
- Then repeat all of the measurements again. If duplicate measurements are not within 1 to 2 mm, measure a third time. Average all measurements from each site.
- For detailed instructions on each site, see "Skinfold Measurement Techniques" on pages 62 and 63.

Use the following formulas for calculating body fat percent:

- **Women (triceps, abdomen, suprailiac)**

  Percent body fat = 0.41563 (sum of 3 skinfolds; triceps, abdomen, suprailiac) − 0.00112 (sum of 3 skinfolds)^2 + 0.03661 (age) + 4.03653

- **Men (triceps, abdomen, suprailiac)**

  Percent body fat = 0.39287 (sum of 3 skinfolds; triceps, abdomen, suprailiac) − 0.00105 (sum of 3 skinfolds)^2 + 0.15772 (age) − 5.18845

- **Women with TRAM-flap reconstruction (triceps, thigh, suprailiac)**

  Db = 1.0994921 − 0.0009929 (sum of 3 skinfolds) + 0.0000023 (sum of 3 skinfolds)^2 − 0.0001392 (age). To calculate percent body fat from Db (body density), use the appropriate formula from table 5.2.

Table 5.3 lists norms for body fat percent.

Pulmonary Function

Forced expiratory volume and forced vital capacity are measured using a dry spirometer. The spirometer printout is placed in the patient's chart. Assessment of pulmonary function is important

Skinfold Measurement Techniques

Triceps

- **Direction of fold:** Vertical (midline).
- **Anatomical reference:** Acromial process of scapula and olecranon process of ulna.
- **Measurement:** Distance between lateral projection of acromial process and inferior margin of olecranon process is measured on lateral aspect of arm with elbow flexed 90°, using a tape measure. Midpoint is marked on lateral side of arm. Fold is lifted 1 cm above marked line on posterior aspect of arm. Caliper is applied at marked level.

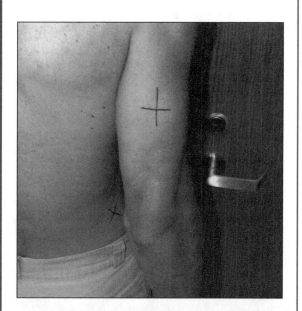

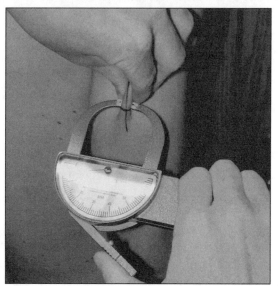

Suprailiac

- **Direction of fold:** Oblique.
- **Anatomical reference:** Iliac crest.
- **Measurement:** Fold is grasped posteriorly to midaxillary line and superiorly to iliac crest along natural cleavage.

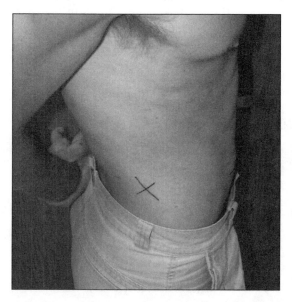

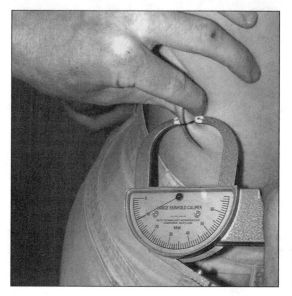

Abdomen

- **Direction of fold:** Horizontal.
- **Anatomical reference:** Umbilicus.
- **Measurement:** Fold is taken 3 cm lateral and 1 cm inferior to center of the umbilicus.

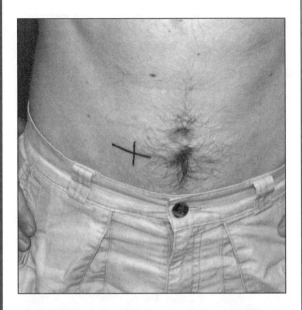

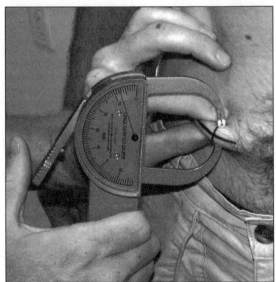

Thigh

- **Direction of fold:** Vertical (midline).
- **Anatomical reference:** Inguinal crease and patella.
- **Measurement:** Fold is lifted on anterior aspect of thigh midway between inguinal crease and proximal border of patella. Body weight is shifted to left foot and caliper is applied 1 cm below fingers.

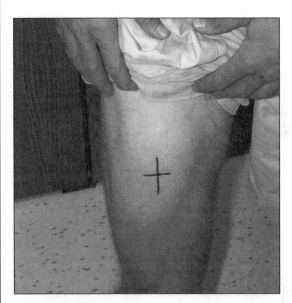

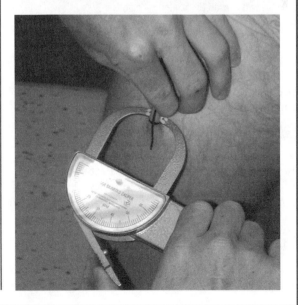

Table 5.2   Population-Specific Conversion of Body Density (Db) to Percent Body Fat

Population	Age	Gender	% Body fat
White	20-80 20-80	Male Female	$(4.95 / Db) - 4.50$ $(5.01 / Db) - 4.57$
Hispanic	20-40	Female	$(4.87 / Db) - 4.41$
Black	18-32 24-79	Male Female	$(4.37 / Db) - 3.93$ $(4.85 / Db) - 4.39$
American Indian	18-60	Female	$(4.81 / Db) - 4.34$

Adapted, by permission, from V.H. Heyward and L.M. Stolarczyk. 1996. *Applied Body Composition Assessment.* (Champaign, IL: Human Kinetics), 12.

Table 5.3   Norms for Body Fat Percent

	Men	Women
At risk[a]	5%	8%
Below average	6%-14%	9%-22%
Average	15%	23%
Above average	16%-24%	24%-31%
At risk[b]	25%	32%

[a]At risk for diseases and disorders associated with malnutrition.

[b]At risk for diseases associated with obesity.

Reprinted, by permission, from V.H. Heyward, 1998, *Advanced Fitness Assessment and Exercise Prescription,* 3rd ed. (Champaign, IL: Human Kinetics), 146.

because of the possible negative side effects of radiation and chemotherapy on pulmonary function as discussed in chapter 2.

Measure forced vital capacity (FVC) and forced expiratory volume in 1 sec (FEV_1). You should obtain two test values. The first value (FEV_1) is the strength of exhalation in the first second, so a forceful initial exhalation is important. The second value (FVC) is the total volume of air the patient can exhale (have the patient continue exhaling as long as possible).

The following standards (norms) are adapted from the American College of Sports Medicine's (ACSM) pulmonary function prediction equations:

95%	Excellent
81-94%	Within normal limits (WNL)
75-80%	Lower limit of normal (LLN)
<75%	Low

Manually calculate percent of predicted FEV_1 and FVC using the following formulas (ACSM, 2000, p. 49):

- **Women**

  FEV_1 (liters) = (0.0268 × ht, cm) − (0.0251 × age) − 0.38

  FVC (liters) = (0.0414 × ht, cm) − (0.0232 × age) − 2.20

- **Men**

  FEV_1 (liters) = (0.0566 × ht, cm) − (0.0233 × age) − 4.91

  FVC (liters) = (0.0774 × ht, cm) − (0.0212 × age) − 7.75

Circumference Measurements

Anthropometric measurements are taken using a calibrated measuring tape. These measurements indicate any circumference changes that result with general edema and with lymphedema. Take measurements on both sides of the body.

For cancer patients, you should not take the measurements over the bulk of a muscle. With this test you are looking for changes over time that may indicate swelling (lymphedema), not changes in muscle mass. You may change the location of the measurement to meet this goal. *Document any changes in landmark location.*

- **Forearm** (3 in. [7.62 cm] up from styloid process of ulna [wrist bone])
- **Upper arm** (5 in. [12.7 cm] up from olecranon process [point of elbow])
- **Lower leg** (5 in. [12.7 cm] up from lateral malleolus [ankle bone])
- **Thigh** (5 in. [12.7 cm] up from superior ridge of patella)
- **Neck** (thyroid cartilage)
- **Subaxillary** (arms abducted 90°)
- **Across nipple** (arms abducted 90°)
- **Beneath breast** (bra line) (arms abducted 90°)

Cardiovascular Endurance

The patient's predicted functional aerobic capacity is determined by a treadmill or bicycle ergometer protocol in the presence of the physician. The physician's presence is necessary in the event of unexpected problems or complications. During the physical examination, the physician has determined the neurological stability of the patient. Chemotherapy can produce neuroendocrine toxicity, which affects motor function. Unstable patients need to have help getting on the bicycle or need to be placed on the bicycle. A patient with motor function difficulties in the lower extremities should use the arm ergometer. If a patient is unable to complete the cardiorespiratory fitness test, a nonexercise physical activity rating can serve as a substitute.

Bruce Treadmill Protocol

Cardiorespiratory assessment using the treadmill follows these guidelines:

- The patient must have a physical exam by the physician before performing this test.
- Use the pulse oximeter throughout the treadmill test.

- Monitor and record blood pressure, heart rate (HR), and rating of perceived exertion during the last minute of each 3-min stage.

The Bruce exercise test, a multistage treadmill protocol (figure 5.1), can be programmed into most treadmill computer systems. Changing both the treadmill speed and the percent grade increases the workload. In the first stage (minutes 1 to 3) of the test, the walking speed is 1.7 mph (45.6 m/min) and the grade is 10%. As the second stage (minutes 4 to 6) begins, increase the grade by 2% and the speed to 2.5 mph (67 m/min). In each subsequent stage, increase the grade 2% and the speed by either 0.8 or 0.9 mph (21.44 or 24.12 m/min) until the subject is exhausted. Prediction equations for this protocol estimate the $\dot{V}O_2$max of active and sedentary women and men.

Record rate of perceived exertion (RPE; figure 5.2) at the end of each stage.

Terminate the test when the client's heart rate reaches 75% of the heart rate reserve (HRR) calculated using the Karvonen formula as follows:

Target HR = (HRmax – HRrest) (.75) + HRrest

Recall that HRmax = 220 – age.

Two other situations warrant test termination. Terminate the test if the patient asks to stop

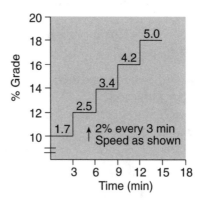

For: normal and high risk
Initial work load: 1.7 mph, 10%, 3 min = normal
1.7 mph, 0-5%, 3 min = high risk

Figure 5.1 The Bruce protocol. This protocol has worked the best at RMCRI; however, any treadmill protocol may be used.

Reprinted, by permission, from R.A. Bruce, F. Kusumi, and D. Hosmer, 1973, "Maximal oxygen intake and nomographic assessment of functional aerobic impairment in cardiovascular disease." *American Heart Journal* 85:546-562.

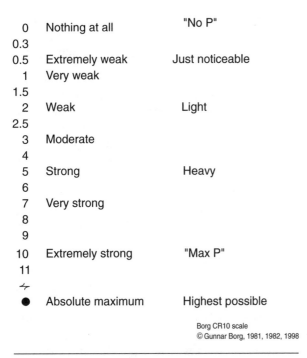

0	Nothing at all	"No P"
0.3		
0.5	Extremely weak	Just noticeable
1	Very weak	
1.5		
2	Weak	Light
2.5		
3	Moderate	
4		
5	Strong	Heavy
6		
7	Very strong	
8		
9		
10	Extremely strong	"Max P"
11		
•	Absolute maximum	Highest possible

Borg CR10 scale
© Gunnar Borg, 1981, 1982, 1998

Figure 5.2 The Borg scale for rating perceived exertion.

Reprinted, by permission from G. Borg, 1998, *Borg's Perceived Exertion and Pain Scales.* (Champaign, IL: Human Kinetics), 47, 50.

because of physical problems such as pain, fatigue, dizziness, or difficulty breathing. You should also terminate the test if the systolic blood pressure drops more than 10 mmHg with an increase in workload or if a hypertensive response occurs (systolic blood pressure >250 mmHg and/or diastolic blood pressure >115 mmHg).

Calculate $\dot{V}O_2$max using one of the following equations.

- **Men and women: used handrails, or cardiac, or elderly**

  $\dot{V}O_2$max = 2.282 (time) + 8.545

  $\dot{V}O_2$max = 2.282 (_____) + 8.545 = _____

- **Women: active or sedentary, no handrails**

  $\dot{V}O_2$max = 4.38 (time) – 3.90

  $\dot{V}O_2$max = 4.38 (_____) – 3.90 = _____

- **Men: active, sedentary, or cardiac, no handrails**

  $\dot{V}O_2$max = 14.76 – 1.379 (time) + 0.451 (time^2) – 0.012 (time^3)

  $\dot{V}O_2$max = 14.76 – 1.379 (_____) + 0.451 (_____) – 0.012 (_____) = _____

$\dot{V}O_2$max norms are displayed in table 5.4.

YMCA Bicycle Protocol

For cardiorespiratory assessment using the bike test (YMCA), monitor the pulse oximeter continuously and follow these guidelines:

- Use a metronome to establish a pedal rate of 50 rpm.
- Initial workload is 150 kgm/min.
- Measure HR during minute 3 of the initial workload to determine the next workload.
- Follow the protocol sheet (figure 5.3) to establish workloads.
- Measure HR during the last 30 sec of minutes 2 and 3 at each workload. If these HRs differ by 5+ bpm, continue the workload for another minute. The patient remains at that workload until a steady state HR is reached or until the HR exceeds 150 bpm.
- Terminate the test when the HR reaches or exceeds 150 bpm.
- Calculate $\dot{V}O_2$max using figure 5.4. Once $\dot{V}O_2$max is determined, compare to norms in table 5.4.

Occasionally a patient is incapable of successfully completing the cardiorespiratory fitness test. In these cases, you can calculate a nonexercise $\dot{V}O_2$max on the basis of the physical activity rating (PA-R). Use the following code to rate the subject's exercise habits:

1. Does not participate regularly in programmed recreation, sport, or physical activity.

   - 0 points: Avoids walking or exertion (always uses elevator, drives instead of walking).

   - 1 point: Walks for pleasure, routinely uses stairs, occasionally exercises sufficiently to cause heavy breathing or perspiration.

2. Participates regularly in recreation or work requiring modest physical activity, such as golf, horseback riding, calisthenics, gymnastics, table tennis, bowling, weightlifting, or yard work.

   - 2 points: 10 to 60 min per week.

   - 3 points: More than 1 hr per week.

3. Participates regularly in heavy physical exercise (running, jogging, swimming, cycling, rowing, skipping rope) or engages in vigorous aerobic activity (tennis, basketball, handball).

Table 5.4 $\dot{V}O_2$max Norms

| | | CARDIORESPIRATORY FITNESS CLASSIFICATION | | | |
| | | MAXIMAL OXYGEN UPTAKE (ML · KG⁻¹ · MIN⁻¹) | | | |
Age	Low*	Fair	Good	Excellent	Superior
Women					
20-29	31	32-34	35-37	38-41	42+
30-39	29	30-32	33-35	36-39	40+
40-49	27	28-30	31-32	33-36	37+
50-59	24	25-27	28-29	30-32	33+
60+	23	24-25	26-27	28-31	32+
Men					
20-29	37	38-41	42-44	45-48	49+
30-39	35	36-39	40-42	43-47	48+
40-49	33	34-37	38-40	41-44	45+
50-59	30	31-34	35-37	38-41	42+
60+	26	27-30	31-34	35-38	39+

*The word "poor" has a negative connotation about the overall health of cancer patients. The word "low" will be used instead of "poor."
The Physical Fitness Specialist Certification Manual, The Cooper Institute for Aerobics Research, Dallas, TX, revised 1997.

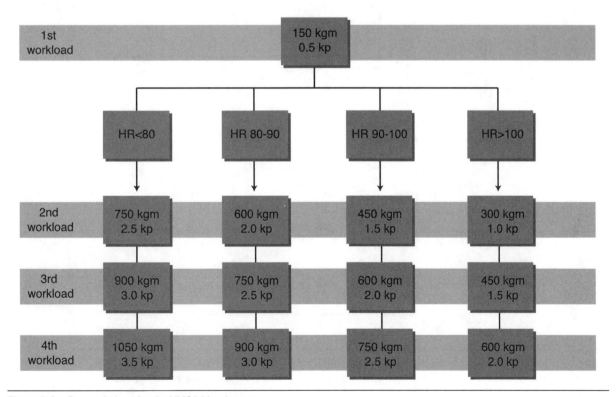

Figure 5.3 Protocol sheet for the YMCA bicycle test.

Adapted, by permission, from L.A. Golding, ed. 2000. *YMCA Fitness Testing and Assessment Manual,* 4th ed. (Champaign, IL: Human Kinetics), 70.

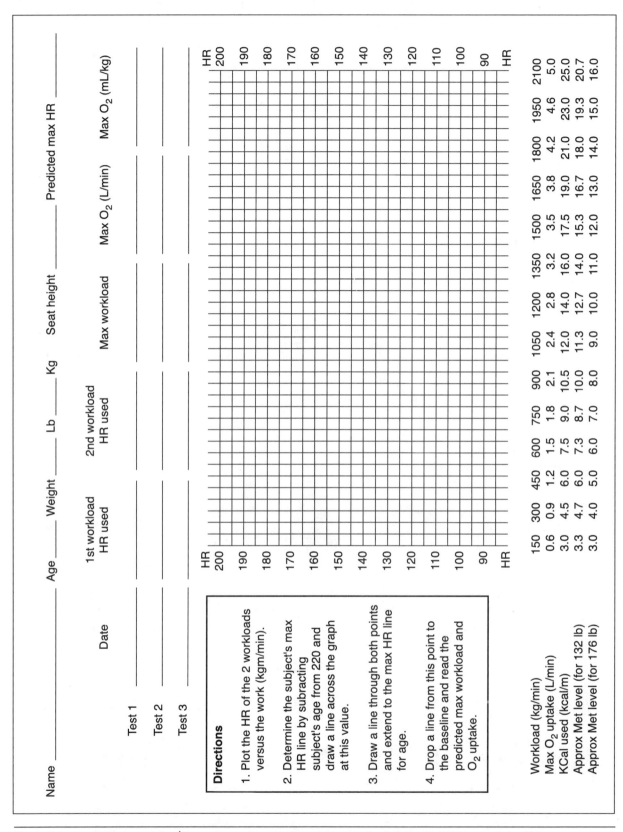

Figure 5.4 Graph for determining V̇O₂max from submaximal heart rates obtained during the YMCA submaximal bike test.

Reprinted, by permission, from L.A. Golding, C.R. Myers, and W.E. Sinning, eds. 1989. *Y's Way to Physical Fitness* (3rd ed.). (Champaign, IL: Human Kinetics), 100.

- 4 points: Runs less than 1 mile (1.61 km) per week or spends less than 30 min per week in comparable physical activity.

- 5 points: Runs 1 to 5 miles (1.61 to 8.05 km) per week or spends 30 to 60 min per week in comparable physical activity.

- 6 points: Runs 5 to 10 miles (8.05 to 16.1 km) per week or spends 1 to 3 hr per week in comparable physical activity.

- 7 points: Runs more than 10 miles (16.1 km) per week or spends more than 3 hr per week in comparable physical activity.

The PA-R is used to estimate $\dot{V}O_2max$ (ml · kg^{-1} · min^{-1}) in the following equation:

% fat model (R = 0.81, SEE = 5.35 ml · kg^{-1} · min^{-1})

$\dot{V}O_2max$ (ml · kg^{-1} · min^{-1}) = 50.513 + 1.589 (PA-R) − 0.289 (age) − 0.552 (%fat) + 5.863 (F = 0, M = 1)

Here is an example of the formula worked out for a 45-year-old female with 25% body fat and a PA-R of 5 (Nieman, 2003):

$\dot{V}O_2max$ = 50.513 + (1.589 × 5) − (0.289 × 45) − (0.552 × 25) + (5.863 × 0) = 31.7 ml · kg^{-1} · min^{-1}

Flexibility and Range of Motion

Flexibility, range of motion, and agility testing procedures incorporate a battery of tests. Have the patient perform a modified sit and reach test (see table 5.5 for norms). Measure shoulder flexion, extension, and abduction and hip flexion and extension range of motion using a portable goniometer and a wall goniometer. The shoulder reach behind back test measures shoulder flexibility. The reason for flexibility and range of motion assessments is that cancer treatments (surgery, radiation, and chemotherapy) alter muscle and connective tissue elasticity and compliance.

Modified Sit and Reach Procedures

For the modified sit and reach test, the patient sits on the floor with shoulders, head, and buttocks against a wall and legs straight out in front. Place a 12-in. (30.5-cm) sit and reach box against the soles of the feet with the zero end of the yardstick toward the patient. Have the patient hold his arms straight forward from the shoulders toward the box, placing one hand on top of the other and keeping the head and shoulders against the wall. Position the yardstick so that the zero end is touching the fingertips. The patient bends forward, sliding

Table 5.5 Norms for Modified Sit and Reach

PERCENTILE RANKS							
		Women			Men		
	Percentile rank	35 years	36-49 years	50 years	35 years	36-49 years	50 years
S	99	19.8	19.8	17.2	24.7	18.9	16.2
	95	18.7	19.2	15.7	19.5	18.2	15.8
E	90	17.9	17.4	15.0	17.9	16.1	15.0
	80	16.7	16.2	14.2	17.0	14.6	13.3
	70	16.2	15.2	13.6	15.8	13.9	12.3
G	60	15.8	14.5	12.3	15.0	13.4	11.5
	50	14.8	13.5	11.1	14.4	12.6	10.2
F	40	14.5	12.8	10.1	13.5	11.6	9.7

(continued)

Table 5.5   *(continued)*

		Women			Men		
		PERCENTILE RANKS					
	Percentile rank	**35 years**	**36-49 years**	**50 years**	**35 years**	**36-49 years**	**50 years**
	30	13.7	12.2	9.2	13.0	10.8	9.3
L	20	12.6	11.0	8.3	11.6	9.9	8.8
	10	10.1	9.7	7.5	9.2	8.3	7.8
VL	05	8.1	8.5	3.7	7.9	7.0	7.2
	01	2.6	2.0	1.5	7.0	5.1	4.0

*Sit and reach scores measured to nearest 0.25 in.

S = superior; E = excellent; G = good; F = fair; L = low; VL = very low.

From *Lifetime Physical Fitness and Wellness,* 5th ed. by Hoeger. © 1998. Reprinted with permission of Wadsworth, a division of Thomson Learning: www.thomsonrights.com. Fax 800-730-2215.

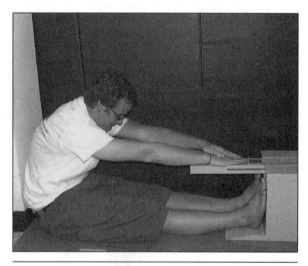

Figure 5.5   The modified sit and reach test.

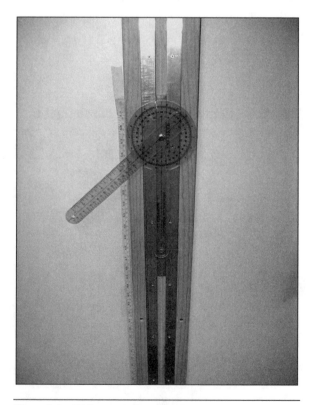

Figure 5.6   Wall-mounted goniometer.

the fingertips along the top of the yardstick (figure 5.5). Read the inches at the farthest tip of the fingertips and record. Compare the reading to the norms in table 5.5.

Goniometer Procedures

At RMCRI, to increase reliability in the use of the goniometer, we mounted it on a piece of wood that we attached to the wall to stabilize one of the arms (figure 5.6). The goniometer can be slid up and down on the piece of wood, and screwed tight at any point. Many patients have difficulty getting onto and off of the floor, so all range of motion (ROM) tests are given in a standing position.

• **Shoulder setting.** Have the patient stand several inches away from the wall; side is to the wall, and one shoulder is in line with the center of the brace. Turn the goniometer knob counterclockwise to loosen; then slide the goniometer so the knob is in line with the acromion process. To tighten, turn the knob *gently* clockwise. Record this setting by lining up the zero on the goniometer with the tape measure on the brace. This setting will be used for all future measurements on the patient.

• **Shoulder flexion.** The palm of the hand faces the body. Keep the elbow straight, and do not allow the patient to twist the shoulders. At maximal flexion, line the moving arm of the goniometer with the line of the humerus. Record.

• **Shoulder backward extension.** The palm of the hand faces the body. Keep the elbow straight, and watch for shoulder rotation and trunk flexion as patient attempts to extend the arm. At maximal extension, line the moving arm of the goniometer with the line of the humerus. Record.

• **Shoulder abduction.** The patient faces the wall with the shoulder joint in line with the knob. The palm of the hand faces anteriorly, causing lateral rotation of the humerus. The elbow must remain straight. At maximal abduction, line the moving arm of the goniometer with the line of the humerus. Record. *Repeat all shoulder tests with the other arm.*

• **Hip setting.** Use the shoulder setting procedures, but line the knob of the goniometer with the lateral aspect of the hip joint (greater trochanter). This landmark is easier to find if you have the patient flex the hip. Record this setting. The patient can place one hand on the wall for support. Use a chair for support on the opposite side if needed.

• **Hip flexion.** Knee flexes during this movement. Watch for trunk flexion. At maximal hip flexion, line the moving arm of the goniometer with the line of the femur. Record.

• **Hip backward extension.** Flex the knee. Watch closely for hip rotation and arching of the back. At maximal hip extension, line the moving arm of the goniometer with the line of the femur. Record. *Repeat all hip tests with the other leg.*

Table 5.6 presents norms for range of motion.

Shoulder Reach Behind Back

Have the patient perform the shoulder reach behind back test for measuring shoulder flexibility (figure 5.7). Slide the hand up the back as far as possible, keeping the wrist straight. Use the middle finger as the pointer. Repeat with the other hand.

Muscular Endurance and Muscular Strength

Patients perform a battery of endurance and strength tests based on varying percentages of their body weight. Tests include bench press, leg press, shoulder press, lateral pull-down, triceps press-down, biceps curl, leg curl, and leg extension. Additionally, the maximum number of curl-ups performed measures abdominal endurance. The handgrip dynamometer measures grip strength. The endurance and strength assessment is necessary because many cancer patients report muscular weakness and fatigue during and following treatment.

Muscular Endurance

Patients execute repetitions until tired for each of the following exercises using a predetermined percentage of their body weight calculated according to their age and sex. The following are important considerations:

1. Ensure that subjects are properly warmed up before initiation of the test protocol.

2. For each individual exercise:

   - Clients should perform as many complete repetitions as possible, but should not be pushed to levels that could induce severe muscle injury.

   - Compute "weight to be lifted" according to the age, sex, and body weight specifications in table 5.7.

   - Assist the client with the first repetition and then continue to spot throughout the test.

   - Repetitions should be performed at a controlled cadence ("1, 2, 3—up"; "1, 2, 3—down").

3. All of the following exercises are specific to Life Fitness exercise equipment. However,

Table 5.6 Norms for Range of Motion

AVERAGE RANGE OF MOTION (ROM) VALUES (IN DEGREES) FOR HEALTHY ADULTS			
Joint	**ROM**	**Joint**	**ROM**
Shoulder		Thoracic-lumbar spine	
Flexion	150-180	Flexion	60-80
Extension	50-60	Extension	20-30
Abduction	180	Lateral flexion	25-35
Medial rotation	70-90	Rotation	30-45
Lateral rotation	90	Hip	
Elbow		Flexion	100-120
Flexion	140-150	Extension	30
Extension	0	Abduction	40-45
Radio-ulnar		Adduction	20-30
Pronation	80	Medial rotation	40-45
Supination	80	Lateral rotation	45-50
Wrist		Knee	
Flexion	60-80	Flexion	135-150
Extension	60-70	Extension	0-10
Radial deviation	20	Ankle	
Ulnar deviation	30	Dorsiflexion	20
Cervical spine		Plantar flexion	40-50
Flexion	45-60	Subtalar	
Extension	45-75	Inversion	30-35
Lateral flexion	45	Eversion	15-20
Rotation	60-80		

Reprinted, by permission, from V.H. Heyward, 2002, *Advanced Fitness Assessment and Exercise Prescription*, 4th ed. (Champaign, IL: Human Kinetics), 234.

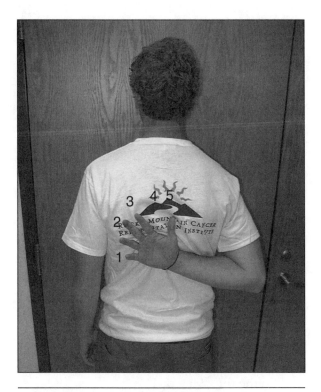

Figure 5.7   Shoulder reach behind back test.

Norms for Shoulder Reach Behind Back Test
5 = excellent
4 = very good
3 = good
2 = fair
1 = poor

any weight equipment can be used with similar modifications.

4. Be sure to remind clients of proper breathing technique during lifting: exhale when lifting; inhale when returning weight.
5. Note that in the exercises with machines, we used LifeFitness Strength equipment (models 8215, multi-press; 8210, lat pull-down/low row; 8245101, leg press/calf; and 24791, leg extension/flexion).

BENCH PRESS

1. Have client lie flat on the bench with feet up on the end of the bench (helps to ensure that back remains flat on the bench throughout the entire test).

2. Client's hands should be placed where they are comfortable, but in a position in which the pectoral muscles are emphasized throughout the lift (hand placement may be dependent on body size).

3. Handles of the bar should be positioned at approximately midchest.

4. Adjust the bar position according to the following procedure:

   - Place the foot (horizontal portion) of the L-square across the shoulder at the olecranon process.

   - The lower edge of the training arm (just above the upper handle) should be 4 to 6 in. (10.16 to 15.24 cm) above the foot of the square.

5. *Full repetition:* UP: Raise training arm until arms are fully extended. DOWN: Lower training arm until weight touches the top of the weight stack.

NOTES

- Add 35 lb (15.89 kg) to amount listed on weight stack to account for the weight of the training arm itself.

- Record training arm setting and hand position for use in reassessment.

LEG PRESS

1. Have the client sit in seat with back flat against backrest and with buttocks tucked as far back as possible.

2. Adjust backrest forward or backward until upper leg is approximately perpendicular to the floor.

3. Have client place feet on the platform approximately shoulder-width apart and in a position in which the quadriceps and hamstrings are exercised as evenly as possible (have the client do a few reps with no weight in order to determine this position).

4. *Full repetition:* UP: Push platform out until legs are fully extended (legs should not be completely locked). DOWN: Allow platform to move back in until weight touches the top of the weight stack.

SHOULDER PRESS

1. Adjust back of seat so that it is in the upright position (make sure that the seat is locked).

2. Have client sit on seat with back and buttocks against the backrest.

3. Client's hands should be placed on the lower handles at a comfortable width (this will depend on body size). Feet should be placed flat on the floor approximately shoulder-width apart.

4. Bar should be set at the lower "shoulder press" setting (second from the top).

5. *Full repetition:* UP: Raise training arm until arms are fully extended. DOWN: Lower training arm until weight touches the top of the weight stack.

LATERAL PULL-DOWN

1. Have client sit on bench seat with thighs positioned comfortably underneath the roller pads (adjust if needed).

2. Client's hands should be positioned on bar slightly outside shoulder-width.

3. Client's torso should remain upright throughout the lift (no leaning forward or backward).

4. *Full repetition:* DOWN: Pull training arm down until it is approximately even with the midchest (pull down in front). UP: Allow training arm to rise until arms are fully extended.

TRICEPS PRESS-DOWN

1. Have client straddle the bench in a standing position with feet approximately shoulder-width apart.

2. Client's hands should be positioned approximately 3 in. (7.62 cm) from the middle of the bar on both sides.

3. Client's torso should remain upright throughout the entire lift (no leaning forward or backward), and the elbows should stay snug to the body.

4. *Full repetition:* DOWN: Push training arm down until arms are fully extended. UP: Allow training arm to rise until it reaches approximately midchest level.

BICEPS CURL (DUMBBELL)

1. Client should be in a standing position with feet approximately shoulder-width apart.

2. Client's torso should remain upright throughout the entire lift (no leaning forward or backward).

3. Client's upper arm and elbow should remain as motionless as possible.

4. *Full repetition:* UP: Raise dumbbell as close to shoulder as possible. DOWN: Lower dumbbell until arm is fully extended.

5. Repeat with other arm.

LEG CURL

1. Have client sit on seat with back and buttocks against backrest, and adjust backrest so that the backs of client's knees are at the forward edge of the seat.

2. Adjust roller pad so that the client's legs start at full extension and the back of the heel rests on the top of the pad.

3. Adjust leg brace so that it rests comfortably on top of the client's thighs.

4. *Full repetition:* DOWN: Pull roller pad down until legs form an angle of 90° (adjust bumper above weight stack so that the weight contacts it at 90° for each repetition). UP: Allow roller pad to rise until weight contacts the stack and legs are fully extended.

LEG EXTENSION

1. Have client sit on seat with back and buttocks firmly against backrest, and adjust backrest so that the backs of the knees are at the forward edge of the seat.

2. Adjust the roller pad so that the client's legs form approximately a 90° angle and the pad rests at the top of the client's feet.

3. Adjust leg brace so that it rests comfortably on top of the client's thighs.

4. *Full repetition:* UP: Raise roller pad up until legs are approximately extended (adjust bumper above weight stack so that the weight contacts it at approximate extension for each repetition). DOWN: Allow roller pad to lower until the weight contacts the weight stack and the legs return to the 90° angle.

The final test for muscular endurance is an abdominal endurance test. If the patient cannot get

Table 5.7 Dynamic Muscular Endurance Test Battery for Cancer Patients of Various Ages*

| | % BODY WEIGHT TO BE LIFTED | | | | | | | |
| Exercise | Age <45 | | Age 45-60 | | Age 60-70 | | Age >70 | |
	Men	Women	Men	Women	Men	Women	Men	Women
Biceps curl L arm	.085	.065	.080	.061	.076	.058	.072	.055
Biceps curl R arm	.085	.065	.080	.061	.076	.058	.072	.055
Bench press	.500	.375	.470	.350	.440	.330	.410	.310
Lat pull-down	.500	.375	.470	.350	.440	.330	.410	.310
Triceps press-down	.250	.250	.230	.230	.210	.210	.190	.190
Leg extension	.375	.375	.350	.350	.330	.330	.310	.310
Leg curl	.375	.375	.350	.350	.330	.330	.310	.310
Leg press	.750	.625	.720	.600	.690	.580	.660	.560
Shoulder press	.300	.225	.280	.210	.265	.200	.250	.185

Record the total number of repetitions completed (max = 15) for all exercises.

Number of repetitions for all exercises will be compared from pre- (initial) to post- (6 months) in order to assess level of improvement.

*The template for this table was originally designed for normal, healthy college-age men and women (shoulder press was added). Two studies that were located suggested that college-age men and women who had survived childhood cancers had approximately 25% less strength than normal, healthy controls of the same age. No other studies comparing muscular strength and endurance in cancer patients and healthy controls could be located. The percentages shown in the original table were reduced by 25% and placed into the first age category (<45). Our age categories were based on research indicating that dynamic and static muscular strength was well preserved until approximately 45 years of age and then declined approximately 5% thereafter. The percentages from the first category (<45) were reduced by 5% for each subsequent age category. This table serves as a "starting point" in view of the fact that research in this area is relatively limited.

onto the floor or has a medical reason why this test cannot be performed, write the reason on the data collection sheet.

CURL-UP (CRUNCH)

1. Have client assume a supine position on a mat with the knees at 90°. The arms are at the side, with fingers touching a piece of masking tape. A second piece of masking tape is placed 8 cm (for those who are >45 years) or 12 cm (for those who are <45 years) beyond the first piece of tape.

2. Set a metronome to 40 bpm and have the client do slow, controlled curl-ups to lift the shoulder blades off the mat (trunk makes a

30° angle with the mat), with fingers touching the second piece of tape, in time with the metronome (20 curl-ups/min). The low back should be flattened before curling up.

3. Client performs as many curl-ups as possible without pausing, up to a maximum of 75. See table 5.8 for curl-up norms.

Muscular Strength

The assessment for muscular strength uses handgrip dynamometer procedures to test grip strength. Before using the handgrip dynamometer, adjust the handgrip size to a position that is comfortable for the individual. Alternatively, you can measure the hand width by means of a caliper and

Table 5.8   Curl-Up (Crunch) Norms

		20-29		30-39		40-49		50-59		60-69	
PERCENTILES BY AGE GROUPS AND GENDER FOR PARTIAL CURL-UP (CRUNCH)											
	Gender	**M**	**F**	**M**	**F**	**M**	**F**	**M**	**F**	**M**	**F**
	Percentile										
Well above average	90	75	70	75	55	75	50	74	48	53	50
	80	56	45	69	43	75	42	60	30	33	30
Above average	70	41	37	46	34	67	33	45	23	26	24
	60	31	32	36	28	51	28	35	16	19	19
Average	50	27	27	31	21	39	25	27	9	16	13
	40	24	21	26	15	31	20	23	2	9	9
Below average	30	20	17	19	12	26	14	19	0	6	3
	20	13	12	13	0	21	5	13	0	0	0
Well below average	10	4	5	0	0	13	0	0	0	0	0

Reprinted, by permission, from American College of Sports Medicine. 2000. *ACSM's Guidelines for Exercise Testing and Prescription* (6th ed.). Philadelphia: Lippincott Williams & Wilkins, p. 86

use this value to set the optimum grip size. The individual stands erect, with arms at the sides. The client should hold the dynamometer parallel to the side, with the dial facing away from the body, and then squeeze the dynamometer as hard as possible without moving the arm. Administer three trials for each hand, allowing a 1-min rest between trials, and use the best score as the client's static strength. Table 5.9 presents dynamometer norms.

Estimated 1-Repetition Maximum for Strength Testing

The 1-repetition maximum (1-RM) is the gold standard for muscular strength testing. This procedure requires lifting the heaviest weight the patient can lift only once, which may not be appropriate for cancer patients. The Rocky Mountain Cancer Rehabilitation Institute is currently developing regression equations that will be used to estimate a 1-RM for muscular strength from the cancer patient's endurance data.

Modified Physiological Assessment Parameters

The need to modify assessments for some patients points to the importance of the physician and the physical examination in the assessment process. Each patient's case differs and requires different modifications; and in some rare cases, the patient cannot do any of the assessments because of poor health. Usually this does not happen, however, and the physician recommends minimal assessments.

The following are abnormalities common in cancer patients, along with typical modifications that are recommended:

• **Hematological abnormalities.** The physician will probably recommend a lower intensity on all the parameters or will eliminate some of the tests, such as the cardiovascular assessment. Again, each patient's case is unique; this is why we recommend individualizing all assessments and prescriptions. As an example, at one time we were drawing blood from patients for research purposes and excluded a patient who had thrombocytopenia because of the physician's instructions. It is difficult to give blood values that definitively contraindicate exercise because many factors are involved, such as age. A chart in chapter 6 (table 6.4) includes blood values from the literature that contraindicate exercise. Be sure you continuously monitor using pulse oximetry and the appearance of the patient.

Table 5.9　Norms for Dynamometer

Classification	<50 years		>50 years	
	Left grip (kg)	Right grip (kg)	Left grip (kg)	Right grip (kg)
Women				
Excellent	>37	>41	>33	>37
Good	34-36	38-40	30-32	34-36
Average	22-33	25-37	19-29	22-33
Low	18-21	22-24	15-18	18-21
Very low	<18	<22	<15	<18
Men				
Excellent	>68	>70	>61	>63
Good	56-67	62-69	49-60	55-62
Average	43-55	48-61	36-48	41-54
Low	39-42	41-47	32-35	34-40
Very low	<39	<41	<32	<34

Adapted, by permission, from V.H. Heyward, 2002, *Advanced Fitness Assessment and Exercise Prescription,* 4th ed. (Champaign, IL: Human Kinetics), 117.

• **Undiagnosed medical conditions.** The physician who performs the physical examination may decide that the patient cannot do the assessments; in these cases we refer the patient back to his own physician for medical clearance. For example, one of our patients had extreme fluctuations in heart rhythm and the physician decided to have her return to her personal physician until she received medical clearance. She returned with a medical clearance after a stress test and we completed the assessment.

• **Neurological abnormalities.** Many of our patients have balance problems due to nerve damage. For these patients the physician always recommends a bicycle ergometer test for the cardiovascular assessment. The physician also recommends spotting the patient on all other assessments to prevent falls. We usually have a second exercise specialist help with these tests. Modified assessments make it possible for patients who cannot get onto the floor to do most of the assessments standing. For example, we have implemented wall push-ups and wall range of motion assessments.

• **Bone pain.** If a patient has bone pain, the physician will recommend substituting light weights for the weight machine portion of the assessment. The cancer exercise specialist needs to be sure to document the assessment procedure so that the same procedure is used for the reassessment.

• **Electrolyte disturbances.** The physician will recommend to patients that they stop any of the assessments immediately and sit down if they are feeling nauseous. Patients occasionally have had to stop the treadmill at a lower intensity (50%) because they were feeling faint. We make sure that patients take in plenty of fluids prior to and during the assessments.

• **Muscle weakness.** Patients who have muscle weakness, depending on its severity, perform modified tests. On the treadmill we modify the heart rate reserve percentage. Weakness in the legs is usually the reason patients cannot do the bicycle test. We modify the strength repetitions on the weight machines or substitute dumbbell exercises. One of our patients was so weak that she could do only the flexibility and range of motion exercises. In her case, the exercise prescription and exercise intervention were based on a goal to complete one repetition or 30 sec on a treadmill. In cases such as this, proper supervision is imperative.

• **Fever.** If patients come in with severely high fevers, the physician usually recommends that they return after the fever is under control and refers them back to their personal physicians.

Subjective Fatigue Assessment

We recommend that clinical settings use both physiological and psychological assessments for cancer patients. The assessments outlined so far evaluate physiological fatigue. A number of established instruments assess subjective fatigue or psychological fatigue perception. Three of the most widely used are the Rhoten Fatigue Scale, the Profile of Mood States, and the Piper Fatigue Scale.

In 1982, the Rhoten Fatigue Scale was developed so that patients could quantify their fatigue (Rhoten, 1982). This scale has an objective observational checklist, which is not consistent with the definition of fatigue as a subjective experience. The Rhoten Fatigue Scale showed a moderate correlation (r = .68) to the fatigue items on the Beck Depression Inventory (Pickard-Holley, 1991). This particular scale takes into account only psychological fatigue and does not include questions concerning physiological fatigue.

The Profile of Mood States (POMS) is used often in cancer research (McNair, Lorr, and Droppleman, 1992). However, Nail (1993) found a pattern of fatigue in cancer patients that differed from the fatigue pattern reflected on the POMS and questioned whether there was a physical component to fatigue, separate from the emotional component. Questioning whether fatigue was physiologically based was important, although Nail did not develop this aspect of the study in much detail.

In 1989, Piper et al. developed the Piper Fatigue Scale (PFS). A total fatigue score is based on four subscales measuring temporal, intensity/severity, affective, and sensory dimensions of fatigue. Internal reliability was above 0.80 in a sample of

breast and lung cancer patients. Although the PFS was useful in that it was designed specifically for cancer fatigue assessment, it was cumbersome and lengthy. In 1998, Piper et al. confirmed the multidimensionality of the PFS and reduced the number of items so that the scale would be more manageable for cancer patients who were experiencing fatigue. Revised following a factor analysis, the PFS now contains four subscales with a total of 22 items. The four multidimensional psychological domains (subscales) are the sensory domain (relating to the sensation of fatigue—whether one feels alert or drowsy, refreshed or exhausted); the affective domain (dealing with the emotional meaning attributed to fatigue ["how you feel about it"]—pleasant or unpleasant, agreeable or disagreeable); the cognitive/mood domain (the degree to which one feels, in relation to fatigue, calm or nervous, happy or sad); and the behavioral domain (the degree to which fatigue interferes with one's ability to do work). These domains contribute to a total fatigue score, which represents total-body fatigue perception in cancer patients following cancer treatments. At RMCRI we use the PFS for subjective fatigue because it is multidimensional. Piper recognized the possibility that fatigue could be both psychological and physiological. The PFS has been used extensively with cancer patients and has excellent *reliability* and *validity*.

Other researchers have developed fatigue scales, but these have not been subjected to extensive psychometric evaluation and do not incorporate the multidimensionality that defines cancer fatigue (Okuyama et al., 2000; Schwartz, 1998; Schwartz and Meek, 1999). Accurate assessment techniques are essential to allow you to evaluate the efficacy of fatigue intervention strategies.

POST-ASSESSMENT PROCEDURES

Following the assessment protocols, the exercise cancer specialist should attend to significant post-assessment activities.

• Be sure all parts of the assessment checklist are completed.

• Record a final heart rate, blood pressure, oxygen saturation, and completion time (ending time of the assessment).

- Set up the next appointment with the patient for the purpose of reviewing and interpreting the assessment results and detailing the individualized exercise prescription that will be developed from those results.

FOLLOW-UP RECOMMENDATIONS

After the patient completes the six-month program, a reassessment determines the effectiveness of the exercise intervention. The postexercise reassessment must be identical to the pre-exercise intervention assessment. A new exercise prescription is developed and a program designed for the next six-month intervention.

SUMMARY

The Rocky Mountain Cancer Rehabilitation Institute adheres to six principles of effective assessment, ensuring that each assessment is relevant to the individual patient and that procedures are reliable and valid, test administration is controlled, patient rights are protected, testing is repeated regularly, and patients understand their assessments. The cancer exercise specialists undergo reliability testing to ensure that their assessments are accurate. A system of physician referral ensures that physicians and cancer exercise specialists work together to provide optimal patient care.

Before coming to the facility for their assessment appointment, patients have filled out a number of forms, such as cancer and medical history forms, a diet analysis, and psychological inventories. To this information are added pre-assessment measures of heart rate, blood pressure, and the like, as well as results of a physical examination by the physician associated with the facility. The patient then undergoes assessment of physiological parameters (body composition, pulmonary function, circumference measurements, cardiovascular endurance, flexibility, and muscle endurance and strength). Assessment procedures are modified for some patients as recommended by the physician who conducted the physical examination. The initial assessment provides baseline data for the development of the individualized exercise program and serves as a basis for comparison with later assessments.

STUDY QUESTIONS

1. Describe the specific criteria that ensure assessment effectiveness.

2. What specific information is essential to establish before you conduct a fitness assessment on a cancer patient?

3. What parameters are important to assess for a cancer patient? Why?

4. How would you ensure the validity and reliability of the protocols used in your assessments?

5. Describe the components of a pre-assessment, an assessment, and the post-assessment process.

6. Identify the contents of a patient chart.

7. Be able to describe in detail how to conduct each assessment procedure (protocols).

8. How would you modify the assessment if the patient had hematological abnormalities and muscle weakness?

9. How would you assess subjective fatigue in a cancer patient?

REFERENCES

American College of Sports Medicine. 2000. *ACSM's guidelines for exercise testing and prescription* (6th ed.). Philadelphia: Lippincott Williams & Wilkins.

Dennehy, C.A., A. Bentz, K. Fox, S.D. Carter, and C.M. Schneider. 1998. Breast cancer risk profile of sedentary, active, and highly trained women. *Med Sci Sports Exerc* 30(5):S63.

Dennehy, C.A., A. Bentz, K. Stephens, S.D. Carter, and C.M. Schneider. 2000. The effect of cancer treatment on fatigue indices. *Med Sci Sports Exerc* 32(5):S234.

Dennehy, C.A., S.D. Carter, and C.M. Schneider. 2000a. Physiological manifestations of prescriptive exercise on cancer treatment-related fatigue. *Physiologist* 43(4):357.

Dennehy, C.A., S.D. Carter, and C.M. Schneider. 2000b. Prescriptive exercise intervention for cancer treatment-related fatigue. *Physical Activity and Cancer: Cooper Aerobics Institute,* No. 4, p. 22.

Dennehy, C.A., C.M. Schneider, S.D. Carter, A. Bentz, K. Stephens, and K. Quick. 2000. Exercise intervention for cancer-related fatigue. *Res Q Exerc Sport* 71(1, suppl):A-27.

Golding, L., Meyers,C., and Sinning, W., eds. 1989. *The Y's way to physical fitness.* Champaign, IL: Human Kinetics.

Heyward, V.H. 1998. *Advanced fitness assessment & exercise prescription* (3rd ed.). Champaign, IL: Human Kinetics.

Heyward, V.H. 2002. *Advanced fitness assessment & exercise prescription* (4th ed.). Champaign, IL: Human Kinetics.

McNair, D., M. Lorr, and L. Droppleman. 1992. *Profile of Mood States manual* (2nd ed.). San Diego: Educational and Industrial Testing Service.

Nail, L.M. 1993. Coping with intracavitary radiation treatment for gynecologic cancer. *Cancer Pract* 1:218–224.

National Institutes of Health; National Heart, Lung, and Blood Institute; National High Blood Pressure Education Program. 1997. The Sixth Report of the Joint National Committee on Prevention, Detection, Evaluation, and Treatment of High Blood Pressure, p. 11. Available online at www. nhlbi.nih.gov /guidelines/hypertension/jncb6.pdf NIH Publication 98-4080.

Nieman, D.C. 2003. *Exercise testing and prescription* (5th ed.). Boston: Mcgraw-Hill, 85-86.

Okuyama, I., T. Akechi, A. Kugaya, H. Okamura, Y. Shima, M. Maruguchi, T. Hosaka, and Y. Uchitomi. 2000. Development and validation of the cancer fatigue scale: A brief, three-dimensional, self-rating scale for assessment of fatigue in cancer. *J Pain Sympt Mgmnt* 19(1):5–14.

Pickard-Holley, S. 1991. Fatigue in cancer patients: A descriptive study. *Cancer Nurs* 14(1):13–19.

Piper, B.F., S.L. Dibble, M.J. Dodd, M.C. Weiss, R.E. Slaughter, and S.M. Paul. 1998. The revised Piper Fatigue Scale: Psychometric evaluation in women with breast cancer. *Oncol Nurs Forum* 25(4):677–684.

Piper, B.F., A.M. Lindsey, M.J. Dodd, S. Ferketich, S.M. Paul, and S. Weller. 1989. The development of an instrument to measure the subjective dimension of fatigue. In *Key aspects of comfort: Management of pain, fatigue, and nausea,* ed. S.G. Funk, E.M. Tournquist, M.T. Champagne, L.A. Copp, and R.A. Weise, 199–208. New York: Springer.

Rhoten, D. 1982. Fatigue and the postsurgical patient. In *Concepts clarification in nursing,* ed. C. Norris, 277–300. Rockville, MD: Aspen.

Schneider, C.M., A. Bentz, and S.D. Carter. 2002a. *The influence of prescriptive exercise rehabilitation on fatigue indices.* Manuscript in preparation.

Schneider, C.M., A. Bentz, and S.D. Carter. 2002b. *Prescriptive exercise rehabilitation adaptations in cancer patients.* Manuscript in preparation.

Schneider, C.M., C.A. Dennehy, M. Roozeboom, and S.D. Carter. 2002. A model program: Exercise intervention for cancer rehabilitation. *J Integr Cancer Ther* 1(1):76–82.

Schneider, C.M., K. Stephens, A. Bentz, K. Quick, S.D. Carter, and C.A. Dennehy. 2000. Prescriptive exercise rehabilitation adaptations in cancer patients. *Med Sci Sports Exerc* 32(5):S234.

Schwartz, A.L. 1998. Reliability and validity of the Schwartz cancer fatigue scale. *Oncol Nurs Forum* 25:711–719.

Schwartz, A.L., and P. Meek. 1999. Additional content validity of the Schwartz cancer fatigue scale. *J Nurs Meas* 7:35–45.

Stephens, K., C.M. Schneider, A. Bentz, M. Lapp, K. Quick, S.D. Carter, and C.A. Dennehy. 2000. The influence of time from cancer treatment on selected fatigue indicators. *Med Sci Sports Exerc* 32(5):S233.

6

Exercise Prescription Development

Each cancer patient presents with a unique combination of cancer type, treatments, medical history, and complications. The exercise prescription must therefore be based on a meticulous examination of the information obtained from the pre-assessment screening and assessment procedures. An individualized exercise prescription for cancer rehabilitation consists of three broad categories of information:

- Health and fitness assessment results
- Exercise recommendations
- Six-month goals

We begin this chapter with a brief discussion of the ways in which assessment affects prescription. We then present exercise recommendations: first, general recommendations for prescribing exercise for cancer patients, and second, recommendations for modifying exercise in response to specific treatment-related symptoms. A final section of the chapter provides information on goal setting and presents sample workouts.

HEALTH AND FITNESS ASSESSMENT RESULTS

The first step in developing an exercise prescription is to record the patient's test results (using the screening, physical examination, and assess-

ment information) and compare these data with norms. Currently, results are compared to normative data for "apparently healthy" individuals because normative data for cancer patients have not yet been developed. (The Rocky Mountain Cancer Rehabilitation Institute [RMCRI] is engaged in developing such data.) An individualized exercise prescription is developed using the data collected and based on the rehabilitation guidelines discussed in chapter 4.

Whatever the assessment results, the total-body fatigue and muscular weakness experienced by most cancer patients dictate that the rehabilitation model will vary from that used for the development of general fitness in an "apparently healthy" population. Regardless of what the assessment shows, the majority of time during the six-month exercise intervention is devoted to whole-body conditioning, whether this is cardiovascular conditioning, muscular conditioning, or a combination of the two. Specificity work is limited and is used only when a particular area of the body is affected by the cancer or cancer treatment (this is discussed at length later in the chapter). It is critical to prescribe resistance exercise for any muscle imbalances revealed by the assessment, as well as to help the patient regain strength lost due to surgery, radiation, or chemotherapy. Ongoing assessment is necessary to enable you to adjust the prescription in response to changes in the patient's condition, whether positive or negative.

In designing the exercise prescription, you will also consider the patient's range of motion (ROM) and balance results. The prescription addresses ROM imbalances uncovered by the assessment and includes stretches for flexibility interspersed throughout the exercise session. It also addresses balance, including the development of proprioceptive abilities *(proprioception)*. The assessment for body composition is necessary to provide a base for monitoring fat and lean mass and should not be used to suggest weight loss, especially to patients who are on chemotherapy. (It is much more important for cancer patients to take in sufficient calories and maintain muscle tissue than to worry about their weight.) The dietary analysis helps you understand how to address the issue of a well-balanced diet with each patient. Finally, you should determine the factors that limit a patient's functional capacity and should design exercise interventions to deal with those weaknesses.

How to respond to specific information provided by assessment is the subject of the remainder of this chapter.

GENERAL EXERCISE RECOMMENDATIONS FOR CANCER REHABILITATION

The components of the exercise prescription for cancer patients are the same as those recommended by the American College of Sports Medicine: mode, frequency, intensity, duration, and progression. Cancer patients, however, have unique needs created by the toxicities that result from the interaction between their cancer and the cancer treatment(s). Exercise can have both positive and negative effects on most of these toxicities. The aim of the exercise guidelines is to preserve the positive relationship between exercise and cancer- and cancer treatment-related symptoms and to avoid any possible negative effect. This section covers the recommendations that apply to all cancer patients, regardless of their type of cancer and treatment regimen. The next section ("Prescription Modifications Addressing Treatment-Related Symptoms," p. 90) shows how to deal with the unique constellation of symptoms that individual patients may have, depending on their type of cancer in combination with the type of treatment.

Carefully adhering to the special considerations for cancer rehabilitation, implementing each of the essential components of the exercise prescription, and utilizing the exercise prescription guidelines thus far developed by RMCRI have resulted in positive improvements in the physiological and psychological health of cancer patients with no adverse effects. The following is a recap of the important points you need to keep in mind in developing the exercise prescription and exercise intervention for cancer patients (Schneider, Dennehy, Roozeboom, and Carter, 2002; Schneider, Bentz, and Carter, 2002a, 2002b).

- Be keenly aware of the special considerations that pertain to cancer rehabilitation patients.
- Use comprehensive physiological and psychological assessments to develop appropriate exercise prescriptions and exercise interventions.
- Base the individualized exercise prescription and intervention program on the type of cancer, the stage of cancer, the severity of treatment, and the time out from treatment.
- Continuously monitor the patient's health status in response to treatment and to the exercise intervention. Change the exercise intervention if the health status of the patient changes.
- The variability that is crucial to effective cancer exercise prescriptions and exercise interventions requires that all exercise specialists in the program be trained in working with cancer patients.
- Base the exercise prescription and exercise intervention program on appropriate types, frequency, intensity, duration, and progression of exercises. Be sure to maintain a logbook of exercise workouts to provide continuity within each patient's program (figure 6.1).

MANIPULATING EXERCISE PARAMETERS TO ESTABLISH EXERCISE DOSES

The health of the immune system is, of course, a major concern for all cancer patients. Exercise dose reportedly influences the immune system. It appears that a moderate exercise dose has a positive effect on the immune system whereas being sedentary and/or exercising intensely has a negative effect (figure 6.2). Adjusting frequency, inten-

DATE:	4/2/02			RHR = 82

Warm-up:	Duration:	Intensity:	RPE:	HR:
Bike	5 min	L1 (Level)	3	91

Aerobic Exercise:	Duration/Mileage:	Intensity:	RPE:	HR:
Bike	5 min	L1	3	90
	5 min	L1	3	95
	3 min	L2	4	106

Resistance Exercise:	Weight:	Sets:	Repetitions:	RPE:	HR:
Leg curl ⑤	40 lb.	2	15	3-4	90
Leg ext. ⑤	45 lb.	2	15	4-5	90
Incline Press ⑤	18 lb.	2	15	3	88
One Arm Bent Over Row ⑤	10 lb.	2	15	3	94
rope Ⓕ		2	10	2	89
Overhead Press ⑤	just	2	15	4	93
handgrips	balls	2	20	3-4	92
Leg lifts (on ball) ⑤		2	15	3	98
Superman (wall)		2	15	3	100
Stretch!					

LIST ANY PROBLEMS OR CONCERNS:
*Legs a little tired today! Do extra stretches!** He's a little tired today, went to hospital last night b/c he was experiencing leakage from his colostomy. Everything is good now.
*Good workout today!

Figure 6.1 Patient's logbook. Warm-up activities, aerobic exercise routine, resistance exercises, and any problems or concerns are identified in detail.

Rocky Mountain Cancer Rehabilitation Institute.

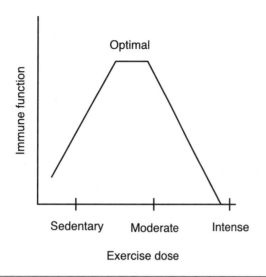

Figure 6.2 Exercise dose and the immune system. According to research on the effects of exercise on the immune system, more is not necessarily better. The exercise prescription should specify moderate levels of exercise for the cancer patient, taking into account the patient's present status and capabilities.

sity, and duration regulates the exercise dose. Exercise intensity appears to be more responsible for reduction in lymphocyte subsets than does the total volume of work (Kajiura, MacDougall, Ernst, and Younglai, 1995). Therefore, the monitoring of lymphocyte subsets from the patient's blood values can help with prescribing the appropriate exercise dose. Laboratory blood values can be obtained from the patient's physicians with the patient's permission.

The following sections present general principles (mode, frequency, intensity, duration, and progression) for adjusting the various parameters of exercise for cancer patients.

Mode

In order to help with the total-body fatigue that occurs in 72% to 95% (National Cancer Institute, 2002) of patients experiencing cancer and cancer treatments, exercises should involve large-muscle groups. Patients should also exercise the specific muscle groups affected by the cancer. The medical director and cancer exercise specialist (and a physical therapist if available) develop this specific work. Specific muscle group work may include range of motion using a ROM wheel for breast cancer patients, physioballs for chemotherapy patients for balance and core strength, and

exerbands or exerbars for resistance training for breast cancer patients and patients who have lost upper body strength due to surgery (Schneider et al., 2002, 2002b). Large-muscle group work can include walking, cycling, wall ball squats for large-muscle groups in the lower body, and swimming (if there is no chance of infection). The elliptical cross-trainer (machine that incorporates a combination of upper and lower body work) is also quite useful for large-muscle work. The mode of exercise will vary among patients. The exercise mode is determined by the health status of the patient, the physiological limits of the patient, and the patient's preferences combined with the exercise goals (table 6.1).

Judy, one of our non-Hodgkin's lymphoma patients on chemotherapy, experienced fatigue and severe balance problems. Judy and her exercise specialist used disc cushions, Aeromat balance pads, and half-round foam rollers to work the whole body to overcome her balance difficulties while also addressing her debilitating fatigue. The exercise specialist chose these particular pieces of equipment because these devices made it easy to work one-on-one with the patient. The specialist needed to be helping the patient at all times, and more advanced pieces of equipment would not have allowed continuous spotting. Also, unlike static balance equipment such as the balance beam, these devices provide multiple proprioceptive stimuli. And finally, besides helping with balance, they are excellent for improving stamina. Since Judy was extremely fatigued and had balance problems, both physiological parameters needed to be improved. Both did improve slightly during her chemotherapy but improved significantly as a result of her six-month posttreatment exercise intervention.

Jim, a patient with advanced prostate cancer, had significant fatigue and muscle wasting with concomitant muscle weakness. A landscape architect, Jim was no longer able to work. Therefore the goal of his exercise program was to improve stamina and muscle strength so he could return to his job. He was not able to use the weight machines because the lightest weight on most of the machines was too heavy for him to lift. He and his exercise specialist decided on wall ball squats using the fit ball to improve his balance, core strength (muscles of the trunk and pelvis that compose the "core" of the body), and full-body strength. Additionally, the exercise intervention included peanut fit balls (balance, core strength, and full-body strength but less intense work), fit tubes for large-muscle upper body (chest and back) work, and body bars for whole-body strength work. The specialist selected these items of equipment because work with them can help with multiple physiological parameters. With these devices, in addition to gaining strength and endurance, Jim was also addressing his whole-body fatigue. Jim experienced increased stamina and strength after one month of exercise. He was able to return to work following his six-month exercise intervention and reported to us that he had no difficulty completing a full day's work.

Susan was a breast cancer survivor who had undergone surgery to remove both breasts. She had recovered her lost stamina but still had minimal ROM in her shoulders. In fact, she could not fix her hair without extreme effort. The cancer exercise specialist determined that the exercise intervention should include very specific work to improve shoulder flexibility. Therefore activities such as the ROM wheel and the pulley ropes were chosen to help with movement in the shoulder and chest area only. Following her six-month program working three days per week for 15 min per day, Susan's shoulder ROM improved to the point that she could fix her hair with minimal effort. Susan told her exercise specialist that the enhanced ROM

significantly improved her quality of life, allowing her to function more efficiently at many daily tasks. Susan's situation is an example of specific muscle work.

Frequency

The exercise prescription should be three days per week, but the frequency will fluctuate depending on treatment status (table 6.2). It may be that a patient is usually able to work three days a week but not during weeks in which she has a chemotherapy or radiation treatment, in which case a two-day intervention may be more appropriate. The frequency will also depend on setbacks or complications during treatment, program goals, duration and intensity of exercise, preferences, time constraints, and functional capacity, as the following examples suggest.

Bob was a 69-year-old terminal lung cancer patient whose exercise frequency was affected by cancer recurrence. He had been coming to RMCRI three days per week and had been doing extremely well in the program. Then his oncologist discovered another spot on his lung. Bob underwent surgery and was not able to return to our program for a few weeks. When he returned, he was very

Table 6.1 Exercise Modes for Cancer Rehabilitation

Fitness component	Exercise modes	Comments
Cardiorespiratory endurance	Walking, jogging, cycling, cross-trainers, swimming (if infection is not possible)	Large-muscle groups Attend to motor function ability Dependent on type of treatment
Muscular strength/endurance	Free weights and machines Resistance balls, resistance bands	Total-body work Weight machine's starting weight is too heavy for most cancer patients
Body composition	Aerobic exercise Resistance exercise	Same as for cardiovascular and muscular strength and endurance
Flexibility	Stretching exercise (static, PNF) ROM wheels, pulleys, flex bands, wall stretching	Attend to surgical and prosthetic areas
Neuromuscular tension/Stress	Progressive relaxation exercise, Tai Chi, movement to music	Depression, anxiety, and stress are prevalent in cancer patients

PNF = proprioceptive neuromuscular facilitation.

weak, and we decided that he should modify the program frequency by exercising twice a day, three days per week, for a shorter duration. Bob lived very near the Institute and was highly motivated to increase his endurance. He was assigned two different trainers, one in the morning and one in the late afternoon. His prescription reflected an initial walking duration of 10 min per session at a low intensity. After three weeks, Bob was able to increase duration to 15 min, and he reported that he was trying to walk every day for at least 10 min. Bob took four months to achieve a walking duration of 20 min, at which time he expressed interest in some weightlifting. Agreeing to continue walking on his own, he returned to a three-day-per-week schedule for 1-hr sessions that included walking and light weightlifting, stretching, and relaxation.

An individual's program goals may also affect frequency, as exemplified by Sharon, a breast cancer patient who was extremely fatigued when she came into our program. One of Sharon's goals was to be able to walk in a Race for the Cure run/walk that would be held four months from when she began her program. She had little support from home and little self-motivation to exercise. Because of her fatigue, her program duration and intensity needed to be low to moderate. To help Sharon meet the goal of walking in the race and given that she lacked motivation to exercise on her own, we decided to establish a frequency of four days per week at our facility instead of three. After one month at RMCRI, Sharon was able to increase her walking duration. Since her goal was to finish the race, we focused on increasing duration, not intensity, and encouraged her to increase frequency by walking on the days of the week she did not come to our program. By race day, Sharon was walking up to 3 miles with her trainer and was successfully walking on her own at least one additional day per week. She did complete the race

and planned to continue her training in hopes of improving her time in the next year's Race for the Cure.

Some of our patients strive to maintain a job during their treatment if at all possible. This imposes time constraints that make it very difficult for them to exercise with us three days per week. We have these patients come in to the program two days a week and give them a home program for the other day, with specific instructions on how to monitor their workouts at home. Because of the demands of work, family, and treatment, the home exercise compliance is very low. As one way to improve compliance in this situation, we strongly encourage family-related physical activity such as biking, walking in the mountains on the weekend, or just a walk in the neighborhood. Involving other family members in this way helped Betty, a breast cancer survivor. Betty worked as a secretary for an insurance company and had two teenage daughters demanding her shuttle service after work. She could barely manage to attend RMCRI two days per week. With persistent encouragement, she finally convinced the family to do a weekly activity at least every other week. In general, we have found that patients who work at home do not show the same degree of improvement as patients who attend the Institute on a regular basis.

Intensity

Intensity should be moderate (table 6.3), the specific level depending on the results of the assessment for health and fitness status. The intensity may range from 30% to 75% of heart rate reserve (Karvonen formula). It is better to err on the side of light to easy intensity than to have the patient work too hard. Karvonen's method for determining heart rate reserve (HRR) (target HR = [(220 − age) − HRrest] × %exercise intensity + HRrest)

Table 6.2 Frequency of Cancer Exercise Sessions

Status	Frequency
Sedentary, poor health and fitness	More than once per day for short bouts
	Minimum of 3 days per week
	Daily exercise to improve health, alternate types of exercise
Active, good health and fitness	2 to 4 days per week to maintain fitness

(Heyward, 2002), discussed on pages 39-40, or Borg's revised rating of perceived exertion (RPE) scale (Borg, 1998) on page 66 is used to assign and monitor intensity. As we will see, in some situations one method is preferable to the other.

The choice of the specific intensity level depends on program goals, age, capabilities (e.g., for a wheelchair user), and health and fitness level. Program goals strongly influenced the intensity of workouts in the case of John, a lung cancer patient. John wanted to work toward the goal of completing stage 3 on the bicycle ergometer cardiovascular test to improve his cardiovascular system. Since the bicycle ergometer protocol increases in intensity at each stage, the cancer exercise specialist modestly increased the intensity of John's workouts over the six-month period.

Variables such as age, capabilities, and health and fitness status often influence the intensity prescribed for patients. Beth, a breast cancer patient, was 86 years old when she entered our program. The exercise prescription specified very low intensities because she was not as stable on her feet as she had been previously. Carla, a patient with brain cancer, used a wheelchair. She performed upper body work with her cancer exercise specialist, utilizing resistance bands with minimal resistance until her strength increased. She then progressed to very light dumbbells. As already noted, establishing cancer patients' health and fitness status also helps determine the exercise prescription. Patients with poor health and low fitness will exercise at a much lower intensity than healthier, fit patients.

Sometimes the Karvonen method is used to assign and monitor intensity levels, and at other times it is necessary to use the Borg scale. As an example, Steve (age 72) entered the program at RMCRI after being diagnosed with stage III testicular cancer. His treatment consisted of surgery to remove the affected testicle followed by a regime of chemotherapy since the cancer had invaded his lymph nodes. He was unable to reach 75% of his HRR during the assessment; in fact, he had to terminate his treadmill assessment test at 50% of his

HRR after 2 min into stage 1 of the Bruce protocol. On the basis of that assessment, the cancer exercise specialist began the aerobic portion of Steve's exercise intervention at 40% of his HRR for 5-min intervals. Steve had a predicted maximal heart rate of 148 bpm with a resting heart rate of 75 bpm. Therefore his target heart rate during his exercise intervention was kept near 104 bpm using the HRR (Karvonen) method. Steve had many setbacks during his chemotherapy, so the exercise specialist modified his workout heart rate prior to each exercise session. Following his chemotherapy, Steve continued in our program for six months. His specialist continued to gradually increase the duration of the workout before increasing the intensity. Steve completed the program working at 75% of his HRR (walking or stationary cycling) for 20-min intervals. Additionally, he gained significant strength working with light body bars throughout the six-month intervention.

Another patient, George, a heavy smoker, was diagnosed with heart disease at the age of 39. He was placed on heart medication (beta blockers) to regulate his heart condition. George continued to smoke and six months after the diagnosis of heart disease was diagnosed with lung cancer. He came to RMCRI during his chemotherapy treatment. With this patient the cancer exercise specialist needed to use the Borg RPE scale instead of heart rate to determine intensity because the beta blocker drug affected his exercise heart rate (heart rate remained depressed instead of increasing with exercise). The RPE values were collected during the assessment on the treadmill. The exercise prescription specified that George would work at an exercise intensity of 2 to 3 on the Borg scale, which was approximately 30% to 40% of his HRR. He was able to maintain between a 2 and a 3 during his exercise workouts while on chemotherapy. During his workout, which consisted of treadmill walking, ball balance work, and wall ball squats using the round fit ball and peanut fit balls, the exercise specialist frequently asked him how he felt. George continued improving following his chemotherapy treatments. When he completed his

Table 6.3   Recommended Intensity Levels

Status	Recommended intensity level
Sedentary, poor health, low fitness	30%-45% HRR (starting); RPE = 1-3
Active, moderate health, average fitness	50%-60% HRR (starting); RPE = 4-5

HRR = heart rate reserve; RPE = rating of perceived exertion.

six-month exercise intervention he was able to work at an intensity of 5, which was approximately 60% of his HRR.

Duration

The duration of the exercise depends on the health and treatment status of the patient, exercise intensity, functional capacity, and program goals. Either continuous exercise or intermittent exercise can be beneficial. Intermittent work is used if the patient performs poorly on the aerobic capacity (treadmill or bicycle) test during the assessment. For example, the cancer exercise specialist would use intermittent exercise for Steve, whom we discussed in the preceding section. Steve was able to complete only 2 min of stage 1 on the Bruce protocol. Therefore during his intervention he needed to work at a low intensity for short periods of time.

During his exercise intervention, Steve worked at 40% of his HRR for 5 min. He would then rest before completing another 5 min of work. During rest intervals, it is good to keep the patient moving. Specificity stretching between exercises helps to minimize soreness as well as allowing for rest. Stretches should extend the areas of the body that were used during the exercise. For example, Steve was doing treadmill work, so he needed to stretch the muscles of the lower body—doing just simple stretches such as placing the hands against the wall and stretching the backs of the calf muscles and hamstrings. Following the rest period, Steve would again begin a 5-min workout. The exercise specialist would continually monitor his heart rate. When the exercise workout became too easy and Steve's heart rate did not get to his target, the exercise specialist increased the duration of the workout. Gradually as the duration increased and became easy, the specialist increased the exercise intensity. Each month the exercise specialist recalculated Steve's target heart rate from the Karvonen formula. By the end of the six-month exercise intervention, he was working continuously at 75% of his HRR for 20-min intervals.

Patients may be able to tolerate from 5 to 60 min of exercise. Duration can vary from one workout session to another depending on the patient's status. Duration can increase every two to three weeks until the patient is able to perform one 30-min bout. For older or less fit individuals, you should increase duration, rather than intensity. Increasing duration instead of intensity helps to minimize the chance of falling in older patients or patients with balance problems. In addition, as mentioned previously, patients cannot usually tol-

erate high intensities—the workout should remain moderate because of the effects on the immune system. The following brief examples suggest how the cancer exercise specialist manipulates the various fitness components.

Deb was a breast cancer patient who developed a brain tumor (noncancerous) one year after her initial diagnosis of breast cancer. She had a stressful job and had high levels of fatigue on certain days. Her cancer exercise specialist needed to decrease the duration of her cardiovascular training, decrease the intensity of strength training, and increase whole-body flexibility and relaxation during each session. If this didn't happen, Deb would experience illness—colds and fever—because of a decrease in immune function. Deb's exercise specialist had to be very aware of her fatigue level at every workout. With the altered workout, Deb was able to keep her immune system functioning well and to decrease her incidence of illness.

Wilma, who also had breast cancer, underwent a hip replacement six months after her breast cancer diagnosis. She began her program at 2 mph, 1% grade for 2 min. She was able to progress to 2.3 mph, 2% grade for 7 min. She was unable to increase speed or grade beyond this level, so duration was increased. Wilma progressed to 15 min, 2.3 mph, 2% grade on the treadmill.

Adelle was diagnosed with breast cancer and had a mastectomy on the left side, which decreased her ROM and decreased the use of the arm. She was unable to reach out and pull the car door closed. Her exercise specialist designed a program consisting of 5-min intervals of work to increase strength and ROM on the left side, including the chest and deltoid muscles. Adelle also performed exercises on the ropes and the ROM wheel during the 5-min workouts. Her cancer exercise specialist increased the duration of the workouts when the exercises became easy for Adelle. In addition to working on ROM in the shoulder, the program stressed flexibility in the entire body, using exercises derived from the exercise assessment ROM testing. Along with relaxation techniques, flexibility was used to "wind down" from the exercise session. Recall that stretching between sets slows clients down and ensures rest periods while still providing some activity. Adelle continued during the six-month exercise intervention to increase the duration of her stretching exercises. She was completing 15-min intervals when she finished her program. She could not believe how effective the exercise routine was and expressed tremendous thankfulness for gaining back the quality of life she had had before her cancer.

Progression

Progression is slow and gradual, especially for very fragile patients. Even minimal exercise during the intervention will help patients progress toward a better quality of life.

Follow the principle of progressive overload, increasing frequency, duration, or intensity gradually, one component at a time. For cancer patients, progression should be much slower than for the healthy population. You will determine progression from the exercise prescription, from the patient's health status, and from the activities completed in the previous workout session. For patients with balance and mobility problems, it is important to increase duration only, since increasing intensity could create safety concerns.

Wayne, 32 years of age, had stomach cancer and was malnourished and underweight.

He had been a runner previously. He began our program by walking at 3.5 mph, 3% grade for 8 min. On one occasion, without the knowledge of his cancer exercise specialist, he ran on his own at home at 4.5 mph for 15 min. He was unable to perform his next scheduled exercise session because of fatigue. He was counseled on the importance of slower progression. At the end of his six-month exercise intervention, Wayne was running at 6.0 mph for 30 min.

To conclude this discussion of exercise parameters and dosage, we refer you to Courneya and Mackey's (2001) list of precautions corresponding to specific complications that you need to take into consideration when prescribing exercise for cancer patients (see table 6.4). Additional precautions identified by the current authors are given throughout this text.

Table 6.4 Precautions When Prescribing Exercise for Cancer Survivors

Complication	Precaution
Hemoglobin level <8.0 g/dl	Avoid activities that require significant oxygen transport (i.e., high intensity).
Absolute neutrophil count $<0.5 \times 10^9$/L	Avoid activities that might increase the risk of bacterial infection (e.g., swimming).
Platelet count $<50 \times 10^9$/L	Avoid activities that increase the risk of bleeding (e.g., contact sports or high-impact exercises).
Fever >38° C	Might indicate systemic infection and should be investigated. Avoid high-intensity exercise.
Fever >40° C	Avoid exercise altogether.
Ataxia/Dizziness/Peripheral sensory neuropathy	Avoid activities that require significant balance and coordination (e.g., treadmill).
Severe cachexia (loss of >35% premorbid weight)	Loss of muscle mass usually limits exercise to mild intensity, depending on the degree of cachexia.
Dyspnea	Investigate etiology. Exercise to tolerance.
Bone pain	Avoid activities that increase risk of fracture (e.g., contact sports or high-impact exercises).
Severe nausea	Investigate etiology. Exercise to tolerance.
Extreme fatigue/muscle weakness	Exercise to tolerance.
Dehydration	Ensure adequate hydration.

Adapted, by permission, from K.S. Courneya, JR. Mackey, and L.W. Jones. 2000. Coping with cancer: Can exercise help? *Physician and Sportsmedicine* 28(5): 49-73.

WRITING THE INITIAL PRESCRIPTION

Each prescription includes the patient's assessment results and should clearly identify all cautions and concerns relevant to the patient, as well as the individualized goals and recommendations for intervention, so that the exercise intervention can be appropriately administered. Figure 6.3, the initial exercise prescription for patient Jane Doe, shows the format for the exercise prescription used at RMCRI. To receive the prescription after having come to the assessment appointment, the patient returns to RMCRI for another visit. The cancer exercise specialist explains the exercise prescription in detail, and of course gives the patient a copy. If the patient is choosing to work at home or in another facility, the exercise prescription is designed for the equipment that is available at these other locations.

The exercise prescription may be modified throughout the six-month exercise intervention if the patient's health status changes. Our exercise specialists record modifications in the patient's logbook (figure 6.1), ensuring that we have the initial exercise prescription and also changes that were made throughout the exercise intervention. Following the six-month exercise intervention the patient has a reassessment, at which time we write a new exercise prescription for the patient to use at another facility or at home.

PRESCRIPTION MODIFICATIONS ADDRESSING TREATMENT-RELATED SYMPTOMS

The type of cancer interacts with the type of treatment to present particular challenges to the patient and the exercise specialist. As noted previously, cancer treatments can be divided into three broad categories: surgery, radiation, and chemotherapy. Because the side effects of these three types of treatment overlap, we begin by examining symptoms that are common to all three types of treatment. We then consider those symptoms common to radiation and chemotherapy, and finally symptoms that are unique to each of two of the forms of treatment (surgery and radiation),

with an additional section on bone marrow transplantation.

Complications Common to Radiation Therapy, Chemotherapy, and Surgery

The side effects common to all three of the major forms of cancer treatment are fatigue and edema. Fatigue as reported by the patient on a given day causes us to modify the prescription, although paradoxically, as we discussed in chapter 2, the exercise also often helps lessen fatigue. Careful monitoring for edema sometimes also leads to an adjustment of the prescription, and here too, exercise can help alleviate the condition.

Fatigue

As already noted, for many patients, extreme fatigue is the most distressing effect of both cancer and its treatment. Although the initial exercise prescription is based on the level of the patient's fatigue, the fatigue level can vary over time, making it necessary to modify the prescription. For example, Deb, the breast cancer patient mentioned earlier, needed modifications in her exercise prescription because her fatigue levels were higher on certain days than on others. If her workouts were not altered, she would experience illnesses such as colds and fever.

Edema

With all three categories of cancer treatment, there is always the possibility of edema (swelling). This is a symptom that you must monitor, and you also must take precautions to eliminate any causes for swelling. Some chemotherapy drugs (e.g., androgens) cause water retention leading to swelling. If the swelling becomes severe, it may be necessary to discontinue the drug. Another factor that contributes to swelling due to surgery and radiation (lymphedema) includes removal of the lymph nodes or radiation injury to the lymph nodes, either of which decreases the flow of lymphatic fluid. Lymphedema may occur following breast cancer surgery or may occur years after surgery or radiation. For example, Ann, a breast cancer patient, had no lymphedema after her breast surgery but 12 years later underwent knee replacement surgery and experienced severe lymphedema in the arm on the side of the breast surgery.

Figure 6.3 Sample initial exercise prescription for Jane Doe, including the three major components: the patient's health and fitness assessment results, exercise recommendations based on those results, and six-month goals.

Patient name: Jane Doe   **Date:**   02-29-01

Age: 51   **RHR:** 91   **BP:** 118/81   **Ht:** 5'6"   **Wt:** 146.5

Piper Fatigue Index score: 3.36   **Beck Depression Inventory score:** 2

Cardiovascular disease risk score: 20 (generally average)

Cancer: Breast cancer   **Location:** right breast

Treatment (surgery): Mastectomy, right breast

Adjuvant Tx: Chemotherapy and radiation

Present treatment status: Chemotherapy starting in one week, radiation following

Cardiovascular Assessment Results

Protocol: Bruce treadmill test

Time completed: 6:17

Predicted $\dot{V}O_2$max: 22.87 (ml · kg^{-1} · min^{-1})

Female aged 50-59

Norms for $\dot{V}O_2$max (ml · kg^{-1} · min^{-1})

Low	Fair	Good	Excellent	Superior
24	25-27	28-29	30-32	33+
22.87				

Body Composition Results

Protocol: Skinfolds (three-site)   **Results:** 27.25%

Protocol: BIA   **Results:** 31.5%

Norms for women all ages

At risk*	Above average	Average	Below average	At risk**
32%	31%-24%	23%	22%-9%	8%
	27.25%			

*At risk for obesity-related disorders such as heart disease

**At risk for malnutrition disorders

Pulmonary Function Results

Protocol: Forced vital capacity (FVC)   **Percent of predicted:** 76

Protocol: Forced expiratory volume (FEV$_1$)   **Percent of predicted:** 74

Norms for FVC (% of predicted)

Low	Low limit of normal	Within normal limits	Excellent
<75%	75%-80%	81%-94%	95%
	76%		

MUSCULAR STRENGTH/ENDURANCE:

Test	Results	Category
Handgrip dynamometer Right (kg)	25	Average
Handgrip dynamometer Left (kg)	17.5	Low
Crunches (maximum)	20	Above average

(continued)

Figure 6.3   *(continued)*

	Weight	Reps
Biceps curl L arm	10 lb	17
Biceps curl R arm	10 lb	20
Bench press	50 lb	10
Shoulder press	35 lb	10
Lat pull-down	50 lb	14
Triceps press-down	35 lb	10
Leg extension	50 lb	10
Leg curl	50 lb	16
Leg press	70 lb	15

RANGE OF MOTION (degrees):

Protocol: Goniometer

	Low	Normal	High
Right shoulder flexion (150-180)		170	
Left shoulder flexion		179	
Right shoulder extension (50-60)		56	
Left shoulder extension		54	
Right shoulder abduction (180)	168		
Left shoulder abduction		180	
Right hip flexion (100-120)		101	
Left hip flexion		111	
Right hip extension (30)	27		
Left hip extension	21		

FLEXIBILITY:

Test	Results	Category
Sit and reach (inches)	14.75	Excellent
Shoulder reach behind back		
Right	4	Very good
Left	3	Good

Medical/Physical concerns:

1. Mastectomy plus removal of 17 right axillary lymph nodes (4+ for cancer)
2. Currently undergoing chemotherapy; aerobic activity may need to be adjusted on chemo therapy days due to increased levels of fatigue
3. Port is in place near the left clavicle
4. Radiation will follow chemotherapy
5. Surgical scarring in right breast area and under right arm
6. Slight shoulder range of motion imbalance

Relevant prescription/OTC drugs:

1. Paxil (30 mg/day)
   Relevant potential side effects: postural hypotension, headache, dizziness, dry mouth
2. Coumadin (2 mg/day)
   Relevant potential side effects: decreased clotting ability, potential for bruising, nausea
3. Lipitor (10 mg/day)
   Relevant potential side effects: back pain, headache, abdominal pain, constipation

EXERCISE REHABILITATION RECOMMENDATIONS

Aerobic Activity

Mode

Walking/Stationary bike.

Jane walks 2-3 times per week. Continue walking!

During rehab, use a combination of treadmill and bike per Jane's interest.

If she is feeling fatigued (chemo days), use the bike.

Frequency

Exercise at least 3 days per week.

Rehab session 2 days per week at the cancer exercise facility.

Try to walk on your own 1-2 days per week.

Intensity

HR should be 40-60% HRR if not fatigued from chemo.

122-138 bpm.

RPE of 2-5 (moderate/strong).

If fatigued, and on chemo days, decrease intensity to an RPE of 2-3 (light).

Duration

10-30 min if not fatigued. Decrease time if needed due to chemo.

Duration can be achieved with a combination of equipment or activities—for example, 10-min bike and 10-min walk (depends on patient interest, which may vary from day to day).

Muscle Strength/Endurance Recommendations

Start with ROM exercises; then progress to additional weight. The first two weeks will be training for proper form while lifting.

(continued)

Figure 6.3 *(continued)*

Start with 1 set of each exercise, 10 repetitions.

RPE: 2-3.

Monitor exercises very closely due to placement of port.

ALL exercises must be spotted.

Be mindful of pain from surgical scarring.

Flexibility and Range of Motion Recommendations

Perform major muscle group stretches upon completion of exercises.

During rehab, use the wheel and rope VERY SLOWLY to increase ROM in the right shoulder region.

Use the wheel and rope very gently prior to weightlifting and upon completion of weightlifting.

There should be NO pain with stretching. Feel a slight pull and hold for about 10 sec.

Do not do any stretches that may pull on the port region.

Take circumference measurements weekly to monitor for swelling.

Perform stretches after each exercise session; hold each stretch 10 to 15 sec initially, breathing throughout the stretch.

TREATMENT GOALS—EXERCISE

Patient name: Jane Doe

Goals are based on the assessment and are effective if no further medical problems arise.

Initial assessment   Date: 02-29-01   **Reassessment**   Date:

Initial prescription   Date: 03-07-01   **Prescription**   Date:

	Results	Norm	6-month goals	Results	Norm	6-month goals
Body fat %	27.25	Above average	26.25			
Endurance/Strength *Cardiovascular* $(ml \cdot kg^{-1} \cdot min^{-1})$ Treadmill time	22.87 6:17	Low	18% ↑ 27.0 7:28			
Pulmonary FVC % predicted	76	Low normal	84			
FEV_1 % predicted	74		80			
Handgrip (kg)	R = 25 L = 17.5	Average Low	Maintain 22			
Crunches (max)	20	Above average	31			
Biceps curl	R = 20 L = 17	10 lb	25 22			

	Results	Norm	6-month goals	Results	Norm	6-month goals
Bench press	10	50 lb	15			
Shoulder press	10	35 lb	15			
Lat pull-down	14	50 lb	19			
Triceps press-down	10	35 lb	15			
Leg extension	10	50 lb	15			
Leg curl	16	50 lb	21			
Leg press	15	70 lb	20			
Flexibility *Sit and reach*	14.75	Excellent	Maintain			
Shoulder reach *behind back*	R = 4 L = 3	Very good Good	Maintain 4			
ROM (degrees) *Shoulder flexion*	R = 170 L = 179	Normal Normal	Maintain			
Shoulder extension	R = 56 L = 54	Normal Normal	Maintain			
Shoulder abduction	R = 168 L = 180	Low Normal	180 Maintain			
Hip flexion	R = 101 L = 111	Normal Normal	Maintain			
Hip extension	R = 27 L = 21	Low Low	30 30			

At RMCRI we do circumference measurements (chapter 5) once a week to monitor swelling. If the cancer exercise specialist believes that the swelling is affecting the patient's movement, or if pain is present, the patient is referred to his primary physician or to our medical director. Additionally, we encourage patients to let us know if they feel heaviness, tingling, numbness, or loss of motor control or pain in the affected area. Identifying any edema or lymphedema early makes treatment easier. A number of measures can reduce the risk for lymphedema: avoiding trauma to the affected side, having blood pressure taken on the unaffected side, avoiding lifting heavy objects (weight training should use moderate loads), avoiding tight jewelry on the affected side, avoiding weight gain, and not smoking or consuming alcohol. Treat-

ing lymphedema usually involves manual lymphatic drainage (massage), the use of compression garments (e.g., sleeves), and exercises. During exercise the patient wears the compression garments so that the exercising muscle will contract against the wrap to increase lymphatic flow. Additionally, increasing ROM and strength in the muscles in the affected area can reduce the risk for the development of lymphedema (McTiernan, Gralow, and Talbott, 2000; Dollinger, Rosenbaum, and Cable, 1997).

Sandy, a breast cancer patient, did not have lymphedema immediately after her mastectomy, so in the initial exercise prescription we developed a light workout program for her. Since there was no lymphedema and therefore she did not have to concentrate on specific arm work, she could do exercises to help with the whole-body fatigue that she was experiencing. Once she had been home for a while, she started having some discomfort when lifting her small child. We altered her exercise prescription to include hand and arm exercises to increase the movement of lymphatic fluid.

Joyce, a breast cancer survivor, had a radical mastectomy followed by radiation therapy. She never experienced lymphedema until six years after her cancer treatments, when she had gallbladder surgery. She came to RMCRI with a severely swollen right arm. In conjunction with a physical therapist trained in lymphedema management, we began a program for Joyce. Before each exercise session we took circumference measurements to determine whether there were any changes in the size of the arm. We began a light exercise program of ROM exercises (wall wheel, rope pulleys) and light strength exercises. Joyce's cancer exercise specialist started her on very light dumbbell weights of 1 to 2 lb that utilized the arm and shoulder muscles. Joyce was receiving manual lymphatic drainage from her physical therapist and wore a compression sleeve. She performed her exercises with the compression sleeve on so that the muscles could contract against the sleeve and force fluid out of the arm. The specialist determined the type of weight and exercise according to Joyce's circumference measurements. If the arm circumference measurements did not increase, the trainer did not modify the program; but if the arm showed more swelling, the trainer would reduce weight and reduce the amount of ROM exercise. After six months, Joyce had minimal swelling but still needed to avoid the risks for the development of lymphedema mentioned earlier.

Complications Common to Radiation Therapy and Chemotherapy

Radiation treatment and chemotherapy can each result in leukopenia; thrombocytopenia; anemia; and nausea, diarrhea, and dehydration. These side effects are not as common with radiation therapy as they are with chemotherapy. The severity of the complications varies depending on the dose of radiation and the dose of chemotherapy, as well as on the type of chemotherapy drug (alkylating agents, antimetabolites, antitumor antibiotics, and

alkaloids) used for treatment. For example, most chemotherapy drugs attack the tumor cell and possibly normal cells that are in the process of cell division; however, alkylating agents attack tumor cells and possibly normal cells whether they are resting or dividing. Therefore the complications described next may be more severe for patients receiving an alkylating chemotherapy drug.

Leukopenia

As noted in chapter 2, leukopenia is a common side effect of radiation and chemotherapy. This means, of course, that the immune system is compromised, leaving the patient vulnerable to infection of all sorts. Because of the compromise to the immune system, the exercise prescription must reflect a very moderate activity level as shown earlier in figure 6.2. But the fact that "moderate exercise" is variable between patients again points to the importance of pre-exercise assessments and individualized exercise prescriptions. The immune system is enhanced with moderate exercise and is suppressed with high-intensity exercise (Kajiura et al., 1995; Nieman, Johanssen, Lee, and Arabatzis, 1990).

How do you know if the immune system is enhanced or suppressed? The only method, at present, is to monitor blood values. Pay close attention to the patient's blood values. If the blood chemistry profile decreases, then seek advice from the medical director or the patient's physician. Cancer patients often have suppressed immune systems; therefore the goal of exercise is to enhance the immune system. Researchers have found that moderate exercise enhances the immune system in healthy individuals. Thus, until positive immune system markers exist, we recommend individualized assessment, prescription, and exercise interventions that do not exceed the "moderate" level of intensity. Since we work very hard at maintaining moderate exercise for each patient, we have not had to have any of our patients stop exercise because of decreasing blood values. The sole exception is in patients currently undergoing chemotherapy. For these patients we have had to modify the exercise intervention by decreasing their workout parameters. We ask a series of questions before every exercise workout to ensure that we know how the patient is feeling. (At RMCRI we are investigating various immune system parameters and oxidative stress markers to try to identify a marker that could indicate enhancement or suppression of the immune system.) Additionally, exercise specialists who are not feeling well should not work with patients. Illness such as colds, flu, and sore throats can easily spread to susceptible patients.

Thrombocytopenia

Thrombocytopenia increases blood clotting time; therefore precautions need to be taken during activities that may produce bleeding, such as blood collections. Patients are also very susceptible to bruising. The best way to tell whether patients have low blood platelets (thrombocytopenia) is to monitor their blood profiles. Since physicians refer patients, it is the physician who determines whether or not the patient can exercise with the current blood profile. Our cancer exercise specialists must know and understand the blood profile values and alert the medical director to any changes. With thrombocytopenia, the exercise specialist must be careful with any exercise involving an object that could bump the patient's skin. For example, with patients using free weights or dumbbells, it is essential to be sure that the patient stops the weight before it touches the body. Again, intense exercise could exacerbate the potential for internal bleeding. Refer to table 6.4 for values that warrant various precautions.

Anemia

Altered hemoglobin concentration is among the numerous alterations in blood parameters that cancer patients may show. Patients who have low hemoglobin concentrations are anemic. They can exercise; but again, they should be careful to do so at a low to moderate intensity so that their feelings of fatigue and tiredness are not exacerbated. For patients with anemia it is also important to be sure to monitor their blood values. Refer to table 6.4 for values.

Mike, a colon cancer patient who underwent surgery for the cancer, had normal hemoglobin values when he first came to RMCRI. About three months into his exercise program he started feeling more tired. His exercise specialist noticed that his hemoglobin values had decreased slightly from the baseline values at the time of the assessment. As the two pursued the possible causes, the patient informed his exercise specialist that he had been bleeding from the rectum for the past two months and did not want to tell his physician because he feared that his cancer had returned. The

cancer exercise specialist immediately referred Mike to his physician. There was no new cancer, and the physician was able to stop the bleeding. Mike returned to our program, and we modified his exercise prescription to less intense work until the hemoglobin values returned to normal.

Nausea, Diarrhea, and Dehydration

Naturally, cancer patients who experience severe nausea and diarrhea during the exercise intervention may need to stop exercise. It is important to ask such patients questions before each session about how they are feeling and to modify the exercise prescription accordingly. Usually the modification is to reduce duration or intensity. Sometimes the patient will just need to go home—in these cases, be sure the person can make it home safely. Dehydration is a problem with all of our patients. We automatically assume dehydration and have water or water bottles available at all times. During rest periods, encourage patients to consume water.

Rita, one of our breast cancer patients, came in one day to exercise and as she started her workout began to feel very nauseated. She had just started taking a depression medication that was not setting well with her. We made sure she was doing better before we let her go home.

Complications That May Arise From Surgery

The issues you should be particularly aware of with patients who have had surgery include susceptibility to infection, pain, body image issues, and internal scarring. Scarring is an inevitable part of healing. Internal scarring can be a late complication from surgery. For example, scarring in the abdomen following surgery for abdominal types of cancers (e.g., cervical cancer) can occasionally lead to a bowel obstruction or cause some other blockage. Additionally, scar tissue can restrict flexibility in the affected area.

Incisions and Sutures

Surgical patients often present with recent incisions and sutures that increase the possibility of their developing infection. Thus it is important to monitor factors within your facility that will influence this possibility. These factors include cleanliness of floors, equipment, locker rooms, pools, spas, and the air filtration system. The initial assessment and exercise prescription will be based on the fact that the cancer patient has an incision and sutures. During the physical examination our medical director checks for ROM and mobility. With surgical patients you should be certain that you do not do any exercise assessment or include any activity in the exercise prescription that will cause unnecessary stress on the incision. Swimming activities of any sort are usually avoided in order to prevent possible infections.

Jane, a breast cancer patient, came to our facility following breast reconstruction after her mastectomy. The muscles in her lower abdomen had been used to reconstruct the breast, and this procedure eliminated the belly button. For research purposes, we assess body composition utilizing a skinfold technique. We had just started our cancer rehabilitation program and were still working out the "kinks." The cancer exercise specialist went into the bathroom with the patient to do an abdominal skinfold. She came out "white as a ghost." "What do I do? The patient doesn't have a belly button." We had failed to select an alternate skinfold site for breast cancer patients who had undergone the transverse rectus abdominis myocutaneous technique (TRAM-flap) for breast reconstruction. We now use an alternate skinfold technique that includes a thigh skinfold instead of the abdominal skinfold for patients with TRAM-flaps.

Postsurgical Pain

Another complication that you must address with surgical patients is the pain they experience with movement. The exercise program must be designed to minimize the patient's pain and maximize the patient's comfort. A possible strategy is to exercise the nonsurgical side until the patient can tolerate exercise movement on the affected side.

Marlene had undergone a bilateral mastectomy with lymph node removal, with a port implanted for administration of chemotherapy drugs. The port needed to be carefully guarded, and avoidance of pain was important. Marlene could perform very light exercise with the upper body. We explained that any time she felt even a minimal amount of pain she was to tell her trainer. To avoid pain, the exercise prescription included light stretching, band, and ball work on the upper body rather than strenuous upper or lower body weightlifting. Lower body work was possible, but

it had to remain moderate so that Marlene's already severe fatigue would not be exacerbated.

Loss of Body Parts

Surgical patients may also experience the loss of body parts, which can affect self-image. It is difficult to deal with the loss of a limb, a breast, a jaw, or a large portion of tissue from any part of the body. The exercise prescription must include exercise that will help the patient overcome the impairment.

Jerry, a lung cancer patient, had undergone removal of his right lung and partial removal of his left lung. He was severely limited in the amount of oxygen that he could consume through his partial lung and therefore had very low energy. The exercise prescription was designed to help him cope with this deficiency. The cardiovascular component of his exercise intervention included intermittent exercise, short duration, low intensity, and rest periods with deep breathing. Jerry's exercises included low intensity exercise with bands, balls, and balance mats. Although Jerry's functional capacity was limited, his cardiovascular endurance did improve significantly throughout the six-month exercise intervention.

Carrie, a breast cancer patient, had had a mastectomy 12 years before she entered our program. After the surgery she had been told not to move her arm on the side of the mastectomy. Twelve years later she could not elevate her arm above her waist. The exercise prescription informed the exercise specialist to work all areas of the body but to pay special attention to exercises that could increase Carrie's ROM on the affected side. The exercise trainer began using the ROM wheel and the pulleys in addition to having Carrie do other arm movements. In time Carrie was able to comb her hair using both arms.

Complications Unique to Radiation Therapy

Patients who have received radiation therapy are likely to present not only with the conditions discussed in the previous section, but also with effects unique to this form of treatment: acute and chronic skin reactions. Skin reactions are described in degrees:

- First-degree skin reaction is characterized by loss of hair in the radiated area.

- Second-degree skin reaction shows skin redness in the radiated area, with inhibition or loss of sweat glands.
- Third-degree skin reaction shows deep redness in the skin, blisters, destruction of the sweat glands, and permanent hair loss.
- Fourth-degree skin reaction shows deep blisters and ulcerations and significant pain.

Skin reactions of any degree can be painful. First-degree reactions do not present any physical barriers to exercise, but reactions of the more serious degrees result in increasingly difficult challenges. Care must be taken during the exercise intervention to reduce infection, eliminate skin irritation, and prevent an excessive rise in body temperature. Because of the risk of infection, the exercise prescription should not suggest swimming for cardiovascular endurance. Sitting on a bicycle can be very uncomfortable for an individual who has had radiation for prostate cancer or cervical cancer. Patients in these groups should instead use the treadmill or do outside walking at a moderate intensity. Exerbands cannot be used if they would touch the radiated skin. To avoid pain, patients should not use a heart rate monitor if the chest area has been radiated. It is important to keep the exercise intensity low to moderate in order to prevent body temperature from rising (remember—the higher the intensity, the higher the body temperature). If the air in your facility is hot and humid, have the patient work in a cooler facility to avoid the discomfort of excessive sweating in the radiated skin area.

Jeannette, who was undergoing radiation for breast cancer, experienced burning in the area where the breast was being radiated. She was unable to wear a bra, so activities were limited to no-bouncing types of activities such as biking. A heart rate monitor was not used during this time; instead, heart rate was determined by palpation. If Jeanette was experiencing pain, all upper body exercises were limited or eliminated until the soreness subsided. She attempted flexibility work but did not perform stretches if doing so caused any pain.

Alan received radiation for a brain tumor and had severe soreness in the ear on the radiation side, so he was unable to wear his hearing aid. Alan and the exercise trainer had to learn to communicate with minimal talking. Alan also experienced disturbances in his balance during and following radiation treatment. Balance activities were increased to retrain the vestibular system.

Complications From Bone Marrow Transplantation

Patients with bone marrow transplantation have very fragile immune status, leaving them vulnerable to infection. Bleeding tendencies and anemia also occur in this group of patients and must be taken into consideration when they are exercising. All these conditions are discussed in the earlier section "Complications Common to Radiation Therapy and Chemotherapy," and the same general guidelines apply to bone marrow transplant patients. It is particularly important to follow these individuals' blood values, especially total leukocyte count, platelet count, and hemoglobin concentration. Anyone working with bone marrow transplantation patients must be free from any illness. Other guidelines include scheduling the patient during a time when she will be the only patient in the facility, cleaning all equipment before the patient arrives, and careful monitoring of fatigue level.

Patty was a breast cancer patient with a stem cell transplant, age 42 years. Patty had worked out during chemotherapy, before her transplant, and was showing improvements. After starting her transplant she experienced a setback and was unable to exercise for two weeks. Her body finally started producing red blood cells, so she resumed exercise with a very low intensity walking and ROM program. She was able to increase at a steady rate until she became ill with a cold. The illness caused a two-week setback. Again she resumed her program at a very low intensity. After finishing the six-month intervention, she exercised on her own and hiked every weekend.

SETTING SIX-MONTH GOALS

Patients are given individualized goals based on their exercise prescriptions. These goals vary considerably for patients undergoing treatment compared with posttreatment patients. Patients undergoing treatment have six-month goals based on values we have obtained thus far on RMCRI patients during treatment. However, cancer exercise specialists do not emphasize these goals because patients feel they need to meet their goals or their cancer will return. Goals give the specialist an idea of how to establish progression. The main goal for patients undergoing treatment is to maintain the quality of life they had prior to the cancer treatment. Patients posttreatment have six-month goals based on the improvements we have seen thus far in our cancer patients at RMCRI.

The following are general principles to adhere to when you are setting six-month goals for patients no longer in treatment.

• Cardiorespiratory endurance. The greatest effects on cardiorespiratory endurance from the six-month intervention program will probably occur during the first six to eight weeks. Patients progress slowly, and you should set realistic goals accordingly. For most posttreatment patients, it is reasonable to expect that in six months they will improve as follows:

Month 1	3%
Month 2	2% per week
Months 3+	1% per week

• Pulmonary function. Unless the patient has setbacks, the pulmonary function goal should place the patient into the next higher category in the normative data table. If the individual is already within normal limits range, set "Maintain" for the goal.

• Flexibility and ROM. Goals for these components should represent the next higher category in the normative data table. If the scores for the modified sit and reach test and the shoulder reach behind back test are already good, or higher, set "Maintain" for the goal. With all "low," "very low," or "poor" ROM scores, the goal should place the individual into the "fair" to "good" range within six months. ROM scores below "average" should have a goal of obtaining average. Scores at average should have "Maintain" as a goal. The goals should also include restoring or maintaining balance between the right and left sides.

• Strength and endurance. Handgrip and crunches should be assigned goals in the next higher category of the normative data table unless the scores are already good and above, respectively, or higher. The other endurance and strength goals are not based on norms. Set moderate goals for gains in these areas, taking into consideration the need for muscular balance.

Normative data tables appear in chapter 5 with each assessment procedure. The norm tables are based on a healthy population, since these are the only data that the literature contains at this time. At RMCRI we are in the process of developing norm

tables for cancer patients for each of the assessment protocols.

General Principles of Workout Sessions

You should, of course, observe all the general principles of safe exercise in every cancer-related exercise session. In fact, because of the fragile condition of many cancer patients, safety precautions are doubly important for their well-being. The principles include the following:

• Warm-up and cool-down. The patient must warm up before exercise, especially flexibility and ROM exercises. The warm-up should consist of 5 to 10 min of light exercises specific to the area being rehabilitated. Following the warm-up, stretching exercises such as static stretching, proprioceptive neuromuscular facilitation (PNF) stretching, ROM wheels, pulleys, flex bands, and wall stretching can be incorporated into the exercise prescription. Visit with the patient concerning any surgical and prosthetic areas in order to avoid injury. Flexibility should be a part of the whole-body workout at every exercise intervention session. It is extremely important to do specificity stretching between exercises to minimize soreness. Each stretch should be held for 10 to 30 sec.

• Balance work. You must ensure that balance deficiencies do not endanger the patient during either cardiovascular or strength-training work. Balance is an issue with many cancer patients. Continuous spotting will eliminate any danger (falling). Additionally, you need to know whether or not the patient has the ability to change body positions, such as the ability to move from the floor to a standing position.

• Teaching proper technique. Be sure that patients use the proper technique with all exercises in order to avoid injury and obtain the maximal benefit possible. For example, be certain to define the appropriate ROM for resistance work or establish steady breathing patterns during resistance work.

• Maintaining hydration. Dehydration is a problem with all cancer treatments, and exercise can cause further dehydration. It is essential to have water available for patients at all times. Encourage patients to drink during rest periods even if they do not feel thirsty.

Sample Workout Sessions

This section presents sample workouts for patients during and after cancer treatment—specifically the workouts that Jane Doe performed as she progressed through and then finished her chemotherapy. Notice in the workouts the fluctuations due to the patient's health status (abnormal blood profile). Notice also the alterations in the workout when imbalances occurred, such as modification of the work on the side that had a chemotherapy port. Appendix B contains photographs and instructions for exercises in the sample workouts.

Sample Workout Sessions for Patient in Treatment

Sample Workouts for a Cancer Patient During Treatment (pp. 102-106) provides selected workouts from a six-month program for a patient who was in the process of treatment. In all of these workouts patient Jane Doe wore a heart rate monitor and used a pulse oximeter. Blood pressure was taken and recorded before and after each workout. Jane is a 51-year-old female with Stage I breast cancer in her right breast. She started her exercise sessions a week after her first chemotherapy treatment. More details about Jane are included in chapter 7.

The patient was monitored prior to each session in order to assess her health and fatigue status from her chemotherapy treatment. The workload (duration and intensity) of the strength and endurance components of each session was adjusted accordingly. Progress was slow because the patient's condition fluctuated with treatment. Frequency, duration, and intensity varied from one session to another depending on her health and cancer status. The exercise specialist monitored form on all exercises, watching for neutral back, soft knees during standing, firm but not tight grip on weights and on rails of treadmill, and proper breathing.

Sample Workout Sessions for Patient Posttreatment

Workouts in the posttreatment period (see Sample Workouts for a Cancer Patient Following Treatment, pp. 107-111) always include warm-up and stretching, the cardiovascular and resistance phase, and the cool-down phase. The patient doing the sample workouts from a six-month program for patients posttreatment had a water bottle available at all times for hydration. She wore a heart rate monitor. She had undergone a reassessment, and the posttreatment workouts addressed imbalances that had been identified. Exercise sessions for patients following treatment continue to emphasize full-body workouts but specifically target weak areas.

Sample Workouts for a Cancer Patient During Treatment

The workout should consist of a warm-up, short stretch, the exercise activity phase (cardiovascular and strength training), and the cool-down/stretching phase. The patient should have a water bottle with her at all times for hydration. The patient will wear a heart rate monitor and use a pulse oximeter, and blood pressure will be taken before and after the workout.

Strength/Endurance

Monitor the patient prior to each session, assessing her health and fatigue status from her chemotherapy treatment. Adjust the workload (duration and intensity) accordingly. Progress slowly because fluctuations will occur with treatment. Frequency, duration, and intensity will vary dependent on the patient's health and cancer status. Monitor form on all exercises: neutral back, soft knees during standing, firm but not tight grip on weights and rails of treadmill, and proper breathing. Have water available for hydration at all times.

40% to 42.5% HRR, 122-124 bpm, 2 days per week for all exercisers

Warm-up

Walking on the treadmill at 1.7 mph at .5% grade for 5 min, RPE: 2

Aerobic activity

Walking on the treadmill at 1.7 mph at 1% grade for 10 min, RPE: 2

Resistance activity 2 days per week

Right side is always worked first unless complications on right side, RPE: 2

- Ropes
  - 10 repetitions, facing the wall, arms moving in the sagittal plane
  - 10 repetitions, facing the wall, arms moving in the frontal plane
  - 1 set each
- Wheel
  - 10 repetitions each side
  - 1 set each
- Bench press
  - Lying on weight bench, patient performs full ROM without weights
  - 10 repetitions
  - 1 set
- Bench press
  - Patient lies on weight bench, uses 5-lb weights; spot closely
  - 10 repetitions
  - 1 set
- One-arm lat row
  - Using an 8-lb weight, have patient use bench or counter to support upper body
  - 10 repetitions
  - 1 set
- Front raises
  - 2-lb weights
  - 10 repetitions
  - 1 set
- Biceps curls
  - 5-lb weights, can be performed unilaterally or bilaterally
  - 10 repetitions
  - 1 set
- Triceps kickbacks
  - 3-lb weights, have patient support upper body on bench or counter
  - 10 repetitions
  - 1 set
- Wall ball squats
  - Monitor knee pain; switch to isolation exercises if painful; no extra weight added; make sure the patient moves the feet out in front of the body to keep knees from going past toes; patient should maintain an upright posture
  - 10 repetitions
  - 1 set
- Standing calf raises
  - Patient stands on floor; have patient come up on toes, then lower down to the floor
  - 10 repetitions
  - 1 set

- Fit-ball pelvic motions
  - Anterior-posterior tilts; lateral movements; circular motions, both directions
  - 10 repetitions each action
  - 1 set
- Crunches
  - Hands crossed over chest
  - 10 repetitions
- Prone hip extensions
  - Patient raises one leg off of the floor while hips maintain contact with the floor
  - 10 repetitions each side
  - 1 set

Cool-down

Walk down to the water fountain to get a cold drink.

Stretches

Assist the patient so that correct positioning on each stretch is attained.

- Hamstrings and quadriceps
- Lats
- Biceps
- Triceps
- Shoulders
- Calves
- Lower back and abdominals
- Neck and chest

MONTH 2 – WEEK 1

40% to 42.5% HRR, 122-124 bpm, 2 days per week

Warm-up

Riding the recumbent bicycle for 5 min at level 1; monitor knee pain; RPE: 2

Aerobic activity

Walking on the treadmill at 2.0 mph at 1% grade for 10 min; monitor heart rate and adjust accordingly; RPE: 2

Resistance activity

Right side is always worked first unless complications on right side, RPE: 2

- Ropes
  - 10 repetitions, facing the wall, arms moving in the sagittal plane
  - 10 repetitions, facing the wall, arms moving in the frontal plane
  - 1 set each
- Wheel
  - 10 repetitions each side
  - 1 set
- Chest press
  - Using a red band, have patient perform a seated one-arm chest press; do not wrap the band around patient's body under her arms because of scarring
  - 10 repetitions each side
  - 1 set
- One-arm lat row
  - Using a green band, have patient rest nonworking arm on counter
  - 10 repetitions
  - 1 set
- Lateral raises
  - 3-lb weights
  - 10 repetitions
  - 1 set
- Biceps curls:
  - Green band, one arm at a time
  - 10 repetitions
  - 1 set
- Lying triceps extension
  - 8-lb weight
  - 10 repetitions
  - 1 set
- Leg press
  - Monitor knee pain, switch to isolation exercises if painful
  - Bar plus 40 lb added
  - 10 repetitions
  - 1 set
- Standing calf raises
  - Standing on the edge of a step with hand support

(continued)

- 10 repetitions
- 1 set

- Fit-ball pelvic motions
  - Anterior-posterior tilts
  - Lateral movements
  - Circular motions, both directions
  - 10 repetitions each action
  - 1 set

- Crunches
  - Hands crossed over chest
  - 10 repetitions
  - 1 set

- Toe taps
  - 10 repetitions each side
  - 1 set

- Prone opposite arm and leg raise

- 10 repetitions each side
- 1 set

Cool-down

Walk down to the water fountain to get a cold drink, including a slow walk in the hall.

Stretches

The patient should be performing most of the stretches on her own with your instructions.

- Hamstrings and quadriceps
- Calves
- Lower back
- Chest and lats
- Biceps and triceps
- Shoulders and abdominals
- Neck

MONTH 4 – WEEK 1

40% to 42.5% HRR, 122-124 bpm, 2 days per week

Warm-up

Walking in the halls for 5 min, RPE: 2; standing stretch

Aerobic activity

Walking on the treadmill at 2.5 mph at 1% grade for 12 min, RPE: 2

Resistance activity

Right side is always worked first unless complications on right side, RPE: 2

- Ropes
  - 10 repetitions, facing wall, arms moving in the sagittal plane
  - 10 repetitions, facing wall, arms moving in the frontal plane
  - 2 sets

- Wheel
  - 10 repetitions each side
  - 2 sets

- Bench press
  - 10-lb weights
  - 10 repetitions
  - 2 sets

- One-arm low rows
  - Green band
  - 10 repetitions
  - 2 sets

- Front-to-side arm raises
  - 3-lb weights, bring the arms up in front and then out to side, back to the front, then down
  - 10 repetitions
  - 2 sets

- One-arm isolation biceps curl
  - 10-lb weight
  - 10 repetitions
  - 2 sets

- One-arm triceps pull-down
  - Green band
  - 10 repetitions
  - 2 sets

- Leg extension machine
  - 40 lb
  - 10 repetitions
  - 2 sets

- Leg curl machine
  - 40 lb
  - 10 repetitions
  - 2 sets
- Standing calf raises
  - Standing on the edge of a step with hand support
  - 5 lb in one hand
  - 10 repetitions
  - 2 sets
- Fit-ball pelvic motions
  - Anterior-posterior tilts
  - Lateral movements
  - Circular motions, both directions
  - 10 repetitions each action
  - 2 sets
- Crunches
  - Hands crossed over chest
  - 10 repetitions
  - 2 sets

- Crossed-knee obliques
  - 10 repetitions each side
  - 2 sets
- Prone opposite arm and leg raise
  - 10 repetitions each side
  - 2 sets

Cool-down

Slow walking for 3 to 5 min

Stretches

The patient should be able to perform the appropriate stretches on her own; continue to monitor her form.

- Hamstrings
- Quadriceps
- Lower back
- Chest
- Lats
- Biceps
- Triceps
- Shoulders
- Abdominals
- Neck

MONTH 6 – WEEK 1

47.5% to 50% HRR, 128-130 bpm, 3 days per week

Warm-up

Recumbent bicycle, level 1, 5 min; monitor knee pain, RPE: 2-3; standing stretches

Aerobic activity

Walking on the treadmill, 3.0 mph, 1% grade for 20 min, RPE: 3

Resistance activity

Work right side first, RPE: 3

- Ropes
  - 12 repetitions, facing wall, arms moving in the sagittal plane
  - 12 repetitions, facing wall, arms moving in the frontal plane
  - 2 sets

- Wheel
  - 12 repetitions each side
  - 2 sets
- Bench press
  - 12 lb
  - 12 repetitions
  - 2 sets
- One-arm lat row
  - 15 lb
  - 12 repetitions
  - 2 sets
- One-arm military press
  - Seated with back support, monitor for pain in scar tissue
  - 5-lb weights
  - 12 repetitions
  - 2 sets

(continued)

- Incline dumbbell curls
  - 10-lb weight
  - 12 repetitions
  - 2 sets
- Triceps kickbacks
  - 5 lb
  - 12 repetitions
  - 2 sets
- Leg press
  - Monitor for knee pain
  - Bar plus 50 lb
  - 12 repetitions
  - 2 sets
- Seated calf raises
  - Machine
  - 20 lb
  - 12 repetitions
  - 2 sets
- Fit-ball pelvic motions
  - Anterior-posterior tilts
  - Lateral movements
  - Circular motions, both directions
  - 12 repetitions each action
  - 2 sets

- Fit-ball back extensions
  - Feet braced against the wall
  - 12 repetitions
  - 2 sets
- Fit-ball crunches
  - 12 repetitions
  - 2 sets
- Outside knee oblique
  - Slow repetitions to one side and fast alternating to each side
  - 12 repetitions to each side
  - 2 sets

Cool-down

Walking to cool down

Stretches

The patient should be able to perform the appropriate stretches on her own; still monitor her form.

- Hamstrings and quadriceps
- Biceps and triceps
- Chest and lats
- Lower back
- Shoulders and abdominals
- Neck

FLEXIBILITY

Stretching should be performed on a daily basis. The body should be warmed up before stretching, so a good time to stretch is immediately after the warm-up or the workout. The stretch at the end of an exercise session is a must! At home a good time to stretch is after a warm bath or shower. The flexibility section of the exercise program should be a relaxing time for the patient. Please take scar tissue into consideration when designing the flexibility program.

Month 1: Teaching phase

- Hold the stretches for 10 to 20 sec.
- Assist the patient so that she is maintaining proper form.
- Perform 2 to 3 sets.

Months 2 to 3

Continue teaching proper technique.

- Have the patient perform the majority of the stretches without your assistance.
- Hold the stretch for 10 to 30 sec.
- Perform 2 to 4 sets.

Months 3 to 6

The patient should be able to perform a full body stretch on her own.

- Perform 2 to 4 sets.
- Increase the holding time for each set until the patient is holding the stretch for 30 sec on the last two sets.

Sample Workouts for a Cancer Patient Following Treatment

MONTH 1 – WEEK 1

The workout for Jane Doe following treatment should consist of a warm-up and stretch, the cardiovascular and resistance phase, and the cool-down phase. Patient should have a water bottle with her at all times for hydration. The patient will wear a heart rate monitor. 42.5% to 45% HRR, 124-126 bpm for all exercises, 2 days per week

Warm-up

Walking on the treadmill at 2.0 mph, 1% grade for 5 min, RPE: 2; standing stretches

Aerobic activity

Walking on the treadmill at 2.5 mph, 1% grade for 10 min, RPE: 2

Resistance activity

Right side is always worked first, RPE: 3

- Ropes
  - 12 repetitions, facing the wall
  - 12 repetitions, facing away from the wall
  - 1 set of each
- Wheel
  - 12 repetitions each side
  - 1 set
- Bench press
  - Lying on weight bench, patient performs full ROM with 8-lb free weights
  - 12 repetitions
  - 1 set
- One-arm lat row
  - 8-lb free weights
  - 12 repetitions
  - 1 set
- Front raises
  - 3-lb free weights
  - 12 repetitions
  - 1 set
- Biceps curls
  - 8-lb free weights, proper back and wrist position highlighted, can be performed unilaterally or bilaterally
  - 12 repetitions
  - 1 set

- Triceps kickback
  - 5-lb free weights
  - 12 repetitions each side
  - 1 set
- Wall ball squats
  - No weight; make sure patient moves feet out in front of body to keep knees from going past toes; patient should maintain an upright posture
  - 12 repetitions
  - 1 set
- Leg extension machine
  - 55 lb
  - 12 repetitions
  - 1 set
- Leg curl machine
  - 55 lb
  - 12 repetitions
  - 1 set
- Standing calf raises
  - Standing on floor, patient should come up on toes and then lower down to floor
  - 12 repetitions
  - 1 set
- Wall pelvic tilts
  - 12 repetitions
  - 1 set
- Fit-ball pelvic motions
  - Anterior-posterior tilts
  - Lateral movements
  - Circular motions, both directions
  - 12 repetitions each action
- Wall stands
  - Develop appropriate posture
  - 1 min

Cool-down

Walk down to the water fountain to get a cold drink.

Stretches

Assist the patient so that correct positioning on each stretch is attained.

(continued)

- Hamstrings and quadriceps
- Calves
- Lower back and abdominals
- Chest
- Lats
- Biceps and triceps
- Shoulders

MONTH 2 – WEEK 1

45% to 47.5% HRR, 126-128 bpm, 2 days per week

Warm-up

Riding the recumbent bicycle for 5 min at level 2, RPE 2-3; standing stretches

Aerobic activity

Walking on the treadmill at 3.0 mph, 2% grade for 15 min; monitor heart rate and adjust accordingly, RPE 2-3

Resistance activity

Right side is always worked first, RPE 3

- Ropes
  - 12 repetitions, facing the wall
  - 12 repetitions, facing away from the wall
  - 2 sets of each
- Wheel
  - 12 repetitions each side
  - 2 sets
- Chest press
  - Using green band, patient performs a seated chest press
  - 12 repetitions
  - 2 sets
- One-arm lat row
  - Green band
  - 12 repetitions
  - 2 sets
- Lateral raises
  - 3-lb free weights
  - 12 repetitions
  - 2 sets
- Biceps curls
  - 5-lb free weights
  - 12 repetitions
  - 2 sets

- Lying triceps extension
  - 8-lb free weights
  - 12 repetitions
  - 2 sets
- Wall ball squats
  - 5-lb free weights in each hand
  - 12 repetitions
  - 2 sets
- Leg extension machine
  - 60 lb
  - 12 repetitions
  - 2 sets
- Leg curl machine
  - 60 lb
  - 12 repetitions
  - 2 sets
- Standing calf raises
  - Standing on the edge of a step
  - 12 repetitions
  - 2 sets
- Fit-ball pelvic motions
  - Anterior-posterior tilts
  - Lateral movements
  - Circular motions, both directions
  - 12 repetitions each action
  - 2 sets
- Floor pelvic tilts
  - 12 repetitions
  - 2 sets
- Crunches
  - Arms crossed over chest
  - 12 repetitions
  - 2 sets

- Prone opposite arm and leg raise
  - 12 repetitions
  - 2 sets

Cool-down

Walk down to the water fountain to get a cold drink, including a slow walk in the hall.

Stretches

The patient should be performing the stretches on her own with your instruction.

- Hamstrings and quadriceps
- Calves
- Lower back and abdominals
- Chest and lats
- Triceps and biceps
- Shoulders

MONTH 4 – WEEK 1

42.5% to 45% HRR, 124-126 bpm, 3 days per week

Warm-up

Walking on the treadmill at 2.0 mph, 1% grade for 5 min, RPE 2; standing stretches

Aerobic activity

Cross-trainer, 17 min, level 1, RPE 2

Resistance activity

RPE 2-3

- Ropes
  - 12 repetitions, facing the wall
  - 12 repetitions, facing away from the wall
  - 2 sets of each
- Wheel
  - 12 repetitions each side
  - 2 sets
- Bench press
  - Free weights, 10-lb dumbbells
  - 12 repetitions
  - 2 sets
- Dumbbell lat rows
  - 12-lb dumbbells
  - 12 repetitions
  - 2 sets
- Front-to-side arm raises
  - 3-lb dumbbells; bring arms up in front, then out to side, back to front, then down
  - 12 repetitions
  - 2 sets

- One-arm isolation curl
  - 10-lb dumbbell
  - 12 repetitions
  - 2 sets
- Triceps kickbacks
  - 8-lb dumbbell
  - 12 repetitions
  - 2 sets
- Wall ball squats
  - 5-lb dumbbells in each hand
  - 12 repetitions
  - 2 sets
- Leg extension machine
  - 60 lb
  - 12 repetitions
  - 2 sets
- Leg curl machine
  - 60 lb
  - 12 repetitions
  - 2 sets
- Standing calf raises
  - Standing on the edge of a step
  - 5 lb in one hand
  - 12 repetitions
  - 2 sets
- Fit-ball pelvic motions
  - Anterior-posterior tilts
  - Lateral movements
  - Circular motions, both directions

(continued)

- 12 repetitions each action
- 2 sets

- Fit-ball back extensions
  - Feet secured by exercise specialist
  - 12 repetitions
  - 2 sets

- Crunches
  - 12 repetitions
  - 2 sets

- Crossed-knee oblique
  - 12 repetitions to each side
  - 2 sets

- Prone opposite arm and leg raise
  - 12 repetitions to each side
  - 2 sets

Cool-down

Slow walking for 3 to 5 min

Stretches

The patient should be able to perform all of the appropriate stretches on her own.

- Hamstrings and quadriceps
- Calves
- Lower back
- Chest
- Lats
- Biceps and triceps
- Shoulders
- Abdominals

MONTH 6 – WEEK 1

47.5% to 50% HRR, 128-130 bpm, 3 days per week

Warm-up

Upright bicycle, level 4, 5 min, RPE 2; standing stretches

Aerobic activity

Cross-trainer, level 3, 22 min, RPE 3

Resistance activity

Right side is always worked first, RPE 2-3

- Ropes
  - 12 repetitions, facing the wall
  - 12 repetitions, facing away from the wall
  - 2 sets of each

- Wheel
  - 12 repetitions each side
  - 2 sets

- Bench press machine
  - Bar (35 lb)
  - 12 repetitions
  - 2 sets

- Low rows
  - 30 lb
  - 12 repetitions
  - 2 sets

- Military press
  - Seated with back support
  - 8-lb free weights
  - 12 repetitions
  - 2 sets

- Incline dumbbell curls
  - 10-lb free weights
  - 12 repetitions
  - 2 sets

- Single triceps press-down
  - Red band
  - 12 repetitions
  - 2 sets

- Leg press
  - 75 lb
  - 12 repetitions
  - 2 sets

- Leg extension machine
  - 60 lb
  - 12 repetitions
  - 2 sets

- Leg curl machine
  - 60 lb
  - 12 repetitions
  - 2 sets

- Seated calf raises
  - 10-lb free weight
  - 12 repetitions
  - 2 sets
- Fit-ball pelvic motions
  - Anterior-posterior tilts
  - Lateral movements
  - Circular motions, both directions
  - 12 repetitions each action
  - 2 sets
- Fit-ball back extensions
  - Feet braced against wall
  - 12 repetitions
  - 2 sets
- Double crunches
  - 12 repetitions
  - 2 sets

- Outside knee oblique
  - Slow repetitions and fast alternating repetitions
  - 12 repetitions to each side
  - 2 sets

Cool-down

Walking to cool down

Stretches

- Hamstrings and quadriceps
- Lower back
- Chest and lats
- Biceps and triceps
- Shoulders
- Abdominals

FLEXIBILITY

Patient should perform stretching on a daily basis. The patient's body should be warmed up before the stretching. Patient should always stretch after every exercise session. At home a good time to stretch is after a warm bath or shower.

Month 1

Month 1 is the teaching phase.

- Hold the stretches for 10 to 12 sec.
- Assist the patient to help achieve proper positioning.
- Perform 2 to 3 sets.

Months 2 to 3

Continue teaching proper technique.

- Have patient perform majority of stretches without your assistance.
- Hold the stretch for 10 to 30 sec.
- Perform 2 to 4 sets.

Months 3 to 6

Patients should be able to perform entire body stretch on her own.

- Perform 2 to 4 sets.
- Increase the amount of stretch progressively for each set.

SUMMARY

An individualized exercise prescription for cancer rehabilitation consists of three broad categories of information: health and fitness assessment results, exercise recommendations, and six-month goals. In general, the elements of an exercise program for a cancer patient are the same as for anyone else—mode, frequency, intensity, duration, and progression. Special considerations apply to cancer patients, however, and the specifics of the exercise prescription, as well as adjustments to the prescription over the course of the program, always flow from the health status, unique circumstances, symptoms, needs, and goals of the individual patient.

The type of cancer interacts with the type of treatment the patient is receiving, continually affecting the patient's capabilities and often necessitating prescription modifications to address treatment-related symptoms. Symptoms common to patients undergoing radiation, chemotherapy, and surgery include fatigue and edema. Effects common to radiation and chemotherapy include leukopenia, thrombocytopenia, anemia, nausea, diarrhea, and dehydration. The program for patients who have had surgery must take into account incisions and sutures, postsurgical pain, or loss of body parts. Radiation and bone marrow transplantation cause other unique complications that affect the exercise prescription. All cancer rehabilitation patients have individualized six-month goals and workout sessions that also differ depending on whether the patient is still undergoing treatment or is finished with treatment.

STUDY QUESTIONS

1. What is an exercise prescription?

2. Why do we write exercise prescriptions?

3. What are the components of the exercise prescription for cancer patients?

4. Why do we list the medical concerns on the exercise prescription?

5. How would you modify the components of the exercise prescription (mode, frequency, intensity, duration, progression) for cancer patients?

6. What principles of exercise physiology are applied to the exercise prescription? Explain.

7. Explain the complications common to radiation therapy, chemotherapy, and surgery. How would you modify the exercise prescription for each of these complications?

8. What is the importance of established goals in the exercise prescription?

9. How do the exercise prescription and the exercise intervention interrelate?

REFERENCES

Borg, G. 1998. *Borg's perceived exertion and pain scales.* Champaign, IL: Human Kinetics.

Courneya, K.S., and J.R. Mackey. 2001. Exercise during and after cancer treatment: Benefits, guidelines, and precautions. *Int Sport Med J* 1(5):1–8.

Courneya, K.S., J.R. Mackey, and L.W. Jones. 2000. Coping with cancer: Can exercise help? *Physician and Sportsmedicine* 28(5):49-73.

Dollinger, M., E.H. Rosenbaum, and G. Cable. 1997. *Everyone's guide to cancer therapy* (3rd ed.). Kansas City, MO: Andrews McMeel.

Heyward, V.H. 2002. *Advanced fitness assessment & exercise prescription* (4th ed.). Champaign, IL: Human Kinetics.

Kajiura, J.S., J.D. MacDougall, P.B. Ernst, and E.V. Younglai. 1995. Immune response to changes in training intensity and volume in runners. *Med Sci Sports Exerc* 27:1111–1117.

McTiernan, A., J. Gralow, and L. Talbott. 2000. *Breast fitness.* New York: St. Martin's Press.

National Cancer Institute. 2002. *Information from PDQ for patients.* Available at http://www.cancer.gov/cancerinfo/pdq/ supportivecare/fatigue/patient/. Accessed September 10, 2002.

Nieman, D.C., L.M. Johanssen, J.W. Lee, and K. Arabatzis. 1990. Infectious episodes in runners before and after the Los Angeles Marathon. *J Sports Med Phys Fitness* 30:316–328.

Schneider, C.M., A. Bentz, and S.D. Carter. 2002a. *The influence of prescriptive exercise rehabilitation on fatigue indices.* Manuscript in preparation.

Schneider, C.M., A. Bentz, and S.D. Carter. 2002b. *Prescriptive exercise rehabilitation adaptations in cancer patients.* Manuscript in preparation.

Schneider, C.M., C.A. Dennehy, M. Roozeboom, and S.D. Carter. 2002. A model program: Exercise intervention for cancer rehabilitation. *J Integr Cancer Ther* 1(1):76–82.

7

Exercise Intervention

You have already learned a great deal about working with cancer patients in exercise programs. You know something of what they are going through, in terms of both their disease and its treatment; you know how they can benefit from exercise; you understand how to assess their status and their needs; you have a good idea of how to prescribe exercise for cancer patients in general, as well as how to individualize those prescriptions for different patients; and you have seen many examples of how to adjust prescriptions for patients in response to their changing conditions. There is more to learn, though, before you are ready to begin applying this understanding in a clinical setting.

This chapter provides information about how to use what you know in practical settings with patients. You will read about working with other medical professionals. You will also learn how to establish positive and productive relationships with your patients; find out about the challenges and opportunities of working with patients in groups versus working with them individually, and in center-based versus home-based programs; learn how to present prescriptions to patients, both orally and in written form; and become aware of gender and age differences in needs. Finally, we discuss putting together exercise programs. We include samples of such programs, two during treatment and the other posttreatment. Studying these programs will give you an idea of what the long-term program looked like for a specific patient going through treatment and following treatment, and of the ups and downs, adjustments,

progress, and benefits she experienced. These programs reflect the principles discussed in previous chapters and give you a general idea of what working with a cancer patient over the long term can look like.

CONDUCTING EXERCISE INTERVENTIONS FOR CANCER PATIENTS

Cancer exercise specialists are trained to provide appropriate exercise interventions for cancer patients. But providing an effective intervention program entails many different kinds of responsibilities. As a cancer specialist you will work with other health care professionals such as physicians, dietitians, and physical therapists; and it is important to form positive working relationships with these individuals. Productive relationships with the patients and their families are another important aspect of an effective program. We have found that it is essential to work individually with patients rather than in groups because all cancer therapies have varying side effects and all patients respond differently to a given therapy. We also recommend an organized center-based exercise intervention—we have not had success with individuals working at home. Presenting the final exercise prescription to the patient is another important responsibility, as is carefully explaining the results of the assessment and the reasons for the exercises prescribed.

Interacting With Other Health Care Professionals

Cancer rehabilitation offers patients help with improving their quality of life. In your work as a cancer exercise specialist, the health care professionals with whom you will have the most important interactions will be physicians. You will have day-to-day contact with the medical director of your program, as well as with the patients' array of physicians. You want your program to be as comprehensive as possible, yet you know your limitations. If the staff in your program are experts in exercise, then ancillary services will be necessary to provide for patients' other needs. You may need the expertise of physical therapists, occupational therapists, massage therapists, dietitians, and psychologists. ("Basic Nutrition Recommendations for Cancer Patients" elaborates on the importance of nutrition, for example.) Probably your facility cannot afford to hire all these professionals but instead will establish a referral system. Be sure that your facility and your ancillary professionals are in constant communication so that working relationships are always good.

Basic Nutrition Recommendations for Cancer Patients

Many organizations publish recommendations for dietary intake and meal planning during and following cancer treatment. For example, the American Cancer Society offers suggestions to cancer patients through the National Cancer Information Center's Web site (www.cancer.org). To deal most effectively with the specific issues related to individual cancer patients during and following treatment, however, it is best to work with a registered dietitian.

Nevertheless, not everyone has access to guidance from a registered dietitian. In these situations, it is helpful for the cancer exercise specialist to reinforce some simple suggestions (American Cancer Society, 2002) about eating.

- Unless otherwise specified by a physician or cancer dietitian, a normal, well-balanced, healthy diet like that outlined in the Food Pyramid is recommended if the patient can tolerate the food.

- Patients should eat meals and snacks with sufficient protein and calories.

- Appetite may be better in the morning—if so, the patient should take advantage of this circumstance and eat more at this time.

- Patients who do not feel well and can eat only one or two foods should stick with these foods until more can be added to the diet. A liquid meal replacement may be a good way for patients to get the calories they need.

- It's important to drink at least six to eight glasses of water per day. Carrying a water bottle helps with fluid intake.

- Patients should avoid eating foods that lack nutritional density, such as foods that are high in calories and contain very low levels of useful nutrients.

- Loss of appetite may make it necessary to consume liquid or powdered meal replacements or more frequent small meals. Soft, cool foods such as yogurt or milkshakes are worth trying.

- Nausea and vomiting can limit food intake. The doctor may prescribe antiemetics (drugs) to control nausea and vomiting. It's best in this situation to emphasize foods that are easy on the stomach, such as toast, crackers, and cream of wheat and to avoid fatty-type foods. A patient who is vomiting should not eat or drink until vomiting is under control. Once the patient can tolerate clear liquids, she can try full-liquid foods such as bouillon, cheese soup, and fruit drinks.

- Patients who experience mucositis (oral cavity complications) should report complications to their physician. Brushing, flossing, and rinsing with normal saline or sodium bicarbonate aid in minimizing the complications.

- Patients who experience diarrhea should replenish lost fluids—eating small amounts of food throughout the day; eating foods that contain plenty of sodium (bouillon) and potassium (bananas); and avoiding greasy, fatty, fried foods.

- Lactose intolerance and other gastrointestinal toxicities may occur after treatments that affect the digestive tract, such as abdominal radiation. The diet needs to contain foods that are low in lactose and foods that are easily digested and absorbed.
- Patients who experience constipation should be sure to drink plenty of fluids. It is important to check with the physician to see if the diet can include increased fiber; in

some types of cancer, high fiber is contraindicated.
- When a patient is not obtaining sufficient calories through meals, nutritional supplements may be necessary; however, patients should consult with their physicians about this.
- Following treatment, the patient should try to return to a diet that is well balanced, containing a variety of healthy foods.

Establishing Positive and Productive Relationships With Patients

Research has shown that even brief psychosocial intervention delivered by caring health care providers who are not trained psychotherapists can significantly enhance the overall psychosocial function of recovering patients, which positively affects mortality (Sotile, 1996). Cancer exercise specialists become very important to cancer patients. The specialist is someone who listens, is positive, and is healthy. Cancer patients do not like to be around other cancer patients continuously. Our patients tell us that it is nice to talk about positive things. So your job and your role by their very nature give you a head start toward establishing a positive and productive relationship with your patients.

Keys to building positive relationships with patients include attentiveness to the patient as an individual and good communication. The best way to establish a positive and productive relationship with a patient is to be caring and attentive. For that hour, you are giving that person your undivided attention. Many researchers (Sotile, 1996) have pointed to the importance of goal setting in providing care to recovering patients. Goal setting that is specific to the needs of the patient is a form of attentiveness. You must consider carefully the specifics of patients' lives in designing goals and evaluating the effects of interventions. The patient's exercise goals should be motivating and literally life altering, and to that end they should have meaning to the patient. One of our patients, Mary, said that not having the energy to pick up her children made her feel like an incompetent mother. We set a goal for her to improve her energy and strength so she could carry her children. Because the goal had personal meaning for her, she was highly motivated and excited about start-

ing the exercise intervention. With patients in treatment, setting a goal such as maintaining their energy level so that they can play with their grandchildren or continue playing in their golf league is highly motivating. (It also helps patients' motivation if they can see the progression toward their goals. Cancer patients are highly motivated, and their adherence to our program is exceptional—the Rocky Mountain Cancer Rehabilitation Institute [RMCRI] has a 94% adherence rate. Of course, the major motivator that patients have in common is that they want to feel good again.)

Communication is an important part of forming a good relationship with a patient. Through communication with the patient before each exercise intervention session, you learn about changes in his status that may necessitate modifications in the day's activities. Before each session, you will want to obtain a brief health and fatigue status report from the patient; you will then adjust that day's activities if current status warrants. Recording the activities and the health and fatigue status from each session in the patient's logbook (figure 6.1) helps you monitor the patient. The following examples suggest how communicating with the patient prior to the exercise session, monitoring every detail of the exercise session, and maintaining the logbook help keep you tuned in to the individual patient's needs.

Michelle, a breast cancer patient, was progressing nicely at each exercise session. But after starting on an antidepressant medication, she came in for her session much more fatigued than she had been on other days. Finding out about Michelle's fatigue and the reason for it enabled the cancer exercise specialist to adjust the workout to accommodate and address the patient's current status. The exercise specialist eliminated the cardiovascular workout, decreased the strength workout, and increased the amount of time for stretching and flexibility exercises.

Besides serving as a means of communication between the exercise specialist and the patient, the logbook helps ensure continuity in work with patients if there are changes in facility personnel. Jim, one of our cancer exercise specialists, decided to leave our facility to continue his education in medical school. The exercise specialist who replaced him, Kathy, was able to learn about the patient's program from the logbook and know exactly how to continue the exercise intervention. Kathy could also learn about the setbacks that the patient had experienced throughout the previous three months. Whether within the context of changes in personnel or not, it is very reassuring to patients that the exercise specialist knows exactly what happened during previous sessions—and the logbook is key to this knowledge.

Even though you are not trained in psychotherapy, your positive relationships with patients will enable you to help them by offering suggestions about coping. Sotile (1996) identifies four strategies that are appropriate to use with patients in rehabilitation:

• First, "think crisis intervention" (p. 64). Crisis intervention approaches are appropriate because cancer patients perceive their situation as a crisis-generating stressor. You can offer patients and their loved ones advice on possible ancillary counseling services, or can "normalize" the patient's situation by saying something like, "Other of our patients who are experiencing your same situation struggle in the ways you are struggling. Let me suggest that you spend a little time with our counselor, who is an expert on helping individuals face stressful situations."

• Another strategy is to provide support. Solicit the patient's concerns, reflect back to the patient an awareness of her concerns, normalize the patient's fears, and offer reassurance.

• The third strategy is to help patients increase their sense of control. Empower patients to be physically active and become truly engaged in their rehabilitation. Reassure them that with work they can regain strength and function in the area that has been affected by cancer.

• The fourth strategy is to educate the patient about what is happening and what will happen during the recovery process.

We might add here that you are almost certain to encounter patients who have undergone surgical removal of body parts or disfigurement such as loss of hair due to cancer treatments. These experiences are stressors that feel particularly

"crisis generating," and the strategies outlined by Sotile may prove especially helpful for patients in such situations.

A good relationship with the cancer exercise specialist, along with the specialist's suggestions for coping, made a big difference to Diane, a cervical cancer patient whom we mentioned in chapter 2. Diane underwent five months of chemotherapy treatment and as a result gained weight and lost her hair. She felt extremely self-conscious about going out in public. Her oncologist suggested that our program might help her physically and socially. When Diane first came to our program she avoided students and kept a hat on to hide her scalp at all times. By visiting with Diane and encouraging her to work out with her hat off, the cancer exercise specialist made her feel comfortable about her hair loss. The specialist pointed out that many individuals, young and old, have no hair, and in fact that college students think it is neat to shave their heads. Diane became very comfortable with her specialist, and this helped her become less self-conscious about her loss of hair and her weight. By the end of the six-month intervention following her treatment she had lost weight and improved her fitness level—and her hair was gradually returning. Diane had a renewed self-image and was grateful to us for helping her improve her quality of life.

Safety Procedures

In any health- and fitness-related facility it is of paramount importance to establish excellent safety procedures. Pre-exercise screening procedures should be followed carefully; standard care requires that pre-participation screening procedures be developed and implemented (Grantham et al., 1998). Medical clearance from either the referring physician or the facility's medical director is necessary for cancer patients. For patient safety, it is essential that all personnel working with cancer patients during a fitness assessment or exercise intervention be aware of the medical and cancer history, all medications being used (prescription or over-the-counter), general dietary patterns, the patient's fitness level, and the expected goals and outcomes for the exercise intervention. For patients in treatment, the cancer exercise specialist should understand the patient's tolerance for each of the assessment procedures. Since many cancer patients are at high risk for infection, it is especially important that all facilities and equipment be kept clean and be well maintained. For patients who have completed treatment, the cancer exercise specialist should also take into consideration the physiological decrements that may persist following specific cancer treatments.

Exercise prescriptions should be carefully constructed from the patient's medical and cancer history, present medications, treatment effects, and present fitness level. Each prescription should clearly identify all cautions and concerns for the patient, as well as the individualized goals and recommendations for intervention, so that the exercise intervention can be appropriately administered. Before each exercise intervention session, the exercise specialist should review the patient's exercise prescription and any notes taken on the previous exercise session. Regularly ask the patient about changes in diet, medications, health status, and tolerance for the previous exercise session. Record any necessary modifications to the intervention in the patient's logbook and subsequently in the patient's chart. Being able to anticipate and recognize problems may prevent injury or setbacks.

All fitness assessment procedures and exercise intervention programs carry a certain risk; however, the risk is heightened in work with cancer patients. The risks are associated with the following:

- Type of treatment the patient has received or is presently receiving
- The present fitness and health status of the individual
- Limitations for mobility
- Drug-related effects (blockers, diuretics, dilators, etc.)
- Nutritional abnormalities
- Weight gain or loss
- Cancer treatment-related symptoms (nausea, dizziness, anemia, total-body fatigue)
- Indwelling chemotherapy ports or catheters
- Vision or hearing loss
- Memory loss, confusion, disorientation
- Balance abnormalities

Working With Individuals Versus Groups

Because cancer patients have unique and variable physical problems, we recommend individual work with an exercise specialist. When this is not possible, then groups of three patients, maybe four, can work together. The group approach is more feasible when the patients have completed treatment and have been exercising for a few months so that they know the exercise routine and there is less chance of injury. The patients still have individualized exercise prescriptions, but the specialist does all similar activities at the same time (e.g., flexibility work).

Note, however, that group work is never appropriate with patients who have difficulties with balance. Patients with balance problems must be spotted at all times. For example, in the case of Beth, a brain cancer patient, severe balance problems developed from her regime of chemotherapy. She always needed to be spotted in all of her activities. For this reason the exercise specialist could not work with other patients at the same time. Beth is not an exception; balance and other neural problems often develop in cancer patients undergoing chemotherapy regardless of what kind of cancer they have. Balance and other neural problems can be reversed. The exercise intervention needs to include extensive work in these areas. Once the patient is no longer at risk for falling, a small-group situation should work.

Center-Based Versus Home-Based Programs

We encourage our patients to complete their exercise intervention at our facility. This gives them one-on-one attention with the specialist. The patient also learns proper form and techniques and can engage in a greater variety of activities—not only because of the equipment available at the facility but also because the exercise specialists have many exercise ideas that patients would not encounter on their own.

In cases in which the patient cannot come to our facility, though, we write the exercise prescription for the activities the individual can do and wants to do at home. We have nursing students visit the home to help the patient with his program and take heart rate and blood pressure assessments. We also ask patients who exercise at home to document the activities that they complete and record how they felt before and after the exercise. We try to explain during the exercise prescription consultation how to complete the exercises properly.

The Rocky Mountain Cancer Rehabilitation Institute does close at certain times during the year, and our exercise specialists write exercise programs for patients to perform at home during those periods. The patient receives logbook sheets to complete and return when RMCRI opens again. It is the patient who makes the decision to work in our facility or at home. If patients have problems that could cause injury during exercise, we encourage them to change their minds and work in our facility. However, a number of our patients live some distance away and cannot come every week.

Presenting Prescriptions to Your Patients

We give our patients a copy of the computerized exercise prescription as presented in chapter 6. At the time of the exercise prescription consultation, the cancer exercise specialist carefully explains each detail of the prescription—the nature of the recommended activity; the reasons for the recommended activity; and the reasons for the particular intensity, duration, and frequency. The specialist also explains how the exercise will help with a particular weakness identified during the assessment. If a patient has low concentrations of hemoglobin, for example, the specialist explains how the prescribed exercise will improve her blood values. Additionally, the specialist goes over the goals of the six-month intervention—but reinforces that they are only goals and that what is important is for the patient to let the specialist know how she feels each session so that the exercise can be modified if necessary. At this time patients also see a nutrition counselor and receive materials on proper nutrition during cancer recovery. We encourage patients to bring a friend or spouse to the consultation to help with remembering the information (some chemotherapy patients have memory loss).

Patients who choose to work at home get a computerized exercise prescription to take with them. For patients who decide to complete their programs at home, a more thorough explanation is necessary, as is a demonstration of the activities. The home exercise prescription is very basic and

Cancer Rehabilitation Versus Cardiac Rehabilitation

Cancer rehabilitation differs from cardiac rehabilitation in a number of important ways. Cardiac disease and cancer affect different systems, and treatments for the two types of diseases differ markedly in their physiological effects. Cardiac patients are much more similar to one another than cancer patients are. Finally, cardiac patients progress through rehabilitation more predictably and consistently than cancer patients usually can.

Cardiac patients have physical problems related to the cardiovascular system, especially the heart. Cancer, on the other hand, may develop in any system of the body. And whereas heart disease does not spread to other systems, cancer can metastasize to tissues far from its site of origin. The development of cardiac disease is more consistent from person to person (e.g., atherosclerosis), whereas cancer can develop as a result of many types of mutated cells or environmental carcinogens. Finally, the causes of cardiac disease are fairly clearly defined (e.g., smoking), while the many causes of cancer are not defined at all.

Another difference that has important implications for rehabilitation is that the treatment for cardiac disease does not cause further toxicities in the body. Cancer treatments cause many different kinds of toxicities that can continue to develop for years.

Cardiac rehabilitation patients have similar physical problems, making group exercise feasible. Cancer patients, as we have stressed throughout this book, present a wide spectrum of physical problems depending on the type of cancer, the type of cancer treatment, and the interaction between the two.

Cardiac patients undergoing rehabilitation usually show progression from one session to the next, so the exercise prescribed can gradually increase in degree of difficulty. Cancer patients have many ups and downs during and following their treatment. Their health status and thus their ability to perform activities vary from one session to the next, so the exercise progression is often not linear.

These contrasts suggest why and how cancer rehabilitation should differ from cardiac rehabilitation. Cancer rehabilitation should be based on the health status of the patient and the person's response to the cancer therapies. Therefore it follows that cancer rehabilitation must be individualized, and furthermore that the exercise specialist needs to be trained specifically for cancer rehabilitation.

is designed from information in the lifestyle questionnaire, which includes information about exercise the patients have completed in the past and equipment they have available to them. Patients can bring a tape recorder to this consultation if they choose. We are developing videotapes and brochures on exercise for cancer patients so that patients will have more information to take with them. This material will include demonstrations of exercises that patients can do at home, emphasizing the proper technique for each exercise and identifying items of "equipment" that most people have in their homes (e.g., cans of varying weights). The brochure will include explanations of how to do exercises, information on the benefits of the exercises, and photos that show the proper exercise technique. These patients also see a nutrition counselor and receive information on proper diets for cancer recovery. As mentioned previously, student nurses help patients with their home-based exercise prescription.

Meeting Gender- and Age-Based Needs

Among patients who come to RMCRI, we have not seen any differences between genders in the response to the various exercises. In the assessment phase, of course, we use calculations specific to the gender. Exercise interventions, however, are not different for males versus females. As we have emphasized, one intervention differs from the next because interventions are based on the results of the individual patient's assessment.

Older cancer patients often find it difficult to get to the floor and get up from the floor. We modify the assessment, the exercise prescription, and the exercise intervention to incorporate no floor work or at least minimal floor work for older patients. The older patients also want to be spotted more and usually do not want to use machines. The difficulty of the exercise intervention is also less than for younger patients. Patients who are younger are more likely to want to work on their own at home since frequently they still have parenting and work-related responsibilities. For these patients it is often difficult to find specific times to come to our facility.

EXERCISE PROGRAMS

The exercise intervention varies considerably between cancer patients during treatment and those who have completed their treatment. Because patients undergoing treatment have many setbacks, they will typically not see a linear progression similar to the development of fitness in a healthy person. Following treatment, cancer patients respond to the exercise program more like a healthy person, although some toxicities from the cancer treatment may occur years later. In any case, the exercise specialists ask their patients a series of questions before each exercise session to assess where they are in terms of fatigue and health. You will notice that the training pyramid for patients during treatment is different from the one for patients following treatment. For patients during treatment, whole-body work is recommended because fatigue and other symptoms occur in the whole body. After treatment, whole-body work should occur each session; but now the specialist can concentrate on specific weaknesses that the patient is experiencing.

The remainder of this chapter presents examples of individualized cancer rehabilitation programs. Jane Doe, the patient whose assessment and prescription information was presented in the preceding chapter, went through her first exercise program and part of her second program while she was undergoing treatment for breast cancer. Reassessments helped the cancer exercise specialist adjust the prescription to provide maximal benefits while taking into account the patient's varying fatigue levels, treatment effects, health status, and preferences. You can see from the program charts that decisions on exercises were made on the basis of Jane's subjective responses and her medical condition.

Initial Six-Month Program During Treatment

An exercise intervention is based on the initial exercise prescription, which is developed from the screening, physical examination, and assessment data (Schneider, Dennehy, Roozeboom, and Carter, 2002; Schneider, Bentz, and Carter, 2002a, 2002b). The model for patients during treatment is based on a rehabilitation rather than a fitness approach. Figure 7.1 depicts such a model, featuring slow progression with continuous monitoring and adjustments if the health status of the patient changes. Because total-body fatigue is prevalent among cancer patients, this phase of the program focuses on the whole body. The goal is to minimize treatment fatigue and other symptoms and to keep the patient active during treatment. The

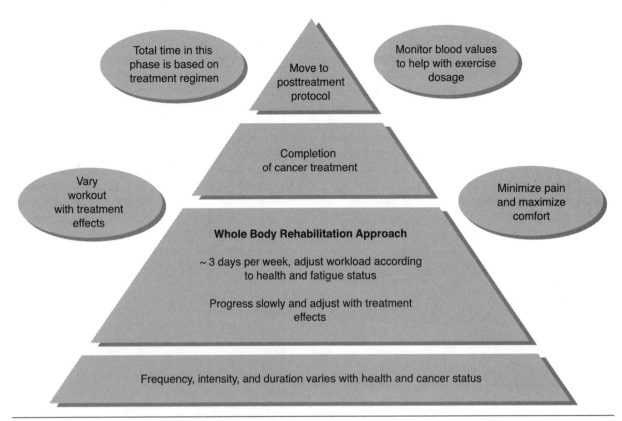

Figure 7.1   Model for six-month exercise intervention during treatment. Central to the intervention during this phase are understanding the effects of the cancer treatment regimen and adjusting the program accordingly.

intervention fluctuates depending on the patient's tolerance for the treatment regimen. An overarching principle is to *emphasize whole-body workouts for every session.* For the patient in treatment, it is vital to stress the importance of remaining active rather than meeting strict goals. Within this model there are no chronological markers—because cancer treatments vary in length and patients respond differently to the treatments, chronological markers are impossible to establish.

Jane Doe was a 51-year-old woman with stage I breast cancer in her right breast. She underwent a mastectomy with some lymph node removal. She was assessed at RMCRI a week before her first chemotherapy treatment and started her exercise sessions a week after the first round of chemotherapy had begun. Chemotherapy consisted of four cycles of Adriamycin and Cytosin, with the cycles three weeks apart. The patient then had weekly treatments of Taxotere, for 12 weeks. Four weeks after the last Taxotere treatment, radiation began; this consisted of a radiation treatment every weekday for seven weeks. The patient had a resting heart rate of 91 (probably elevated due to treatments), with no restrictions on exercise. She

had some scarring under the right arm and a chemotherapy port in her upper chest on the left side. She had led an active lifestyle up to the time of the mastectomy.

Throughout the exercise intervention, the exercise specialist monitored the patient prior to each session, assessing her health and fatigue status from her chemotherapy or radiation treatments and adjusting the workload accordingly. The progression was slow because of the fluctuations that occurred with treatment. Frequency, duration, and intensity varied according to the patient's health and cancer status. All sessions during this initial six-month intervention emphasized whole-body workouts.

The chart "Exercise Program for Jane Doe During Chemotherapy Treatment" details week by week the exercises that the patient performed, as well as her health status. Notice how the program was continually adapted on the basis of new medical information, the patient's subjective responses and perceptions, and her preferences. This highly individualized program reflects the thoughtful engagement of the patient as much as the attentiveness of the exercise specialist.

Exercise Program for Jane Doe During Chemotherapy Treatment

MONTH ONE			
Week	**Aerobic**	**Strength/Endurance**	**Health Status**
1	40%-42.5% HRR 122-124 bpm RPE: 2 2 days per week 10 minutes duration	ROM, light free weights, light resistance bands 2 days per week 1 set 10 repetitions RPE: 2	Patient is in the second week of the first round of chemotherapy. Chemotherapy will be administered 4 times with 3 weeks between doses. She is starting to feel some fatigue with a little nausea. Chemotherapy will be administered on Monday; the client has chosen to work out on Tuesday and Thursday.
2	40%-42.5% HRR 122-124 bpm RPE: 2 2 days per week 10 minutes duration	ROM, light free weights, light resistance bands 2 days per week 1 set 10 repetitions RPE: 2	Patient was fatigued but feeling better than last week. She felt she was able to perform the same activities as the week before.
3	40%-42.5% HRR 122-124 bpm RPE: 2 2 days per week 10 minutes duration	ROM, light free weights, light resistance bands 2 days per week 1 set 10 repetitions RPE: 2	Patient underwent second round of chemotherapy on Monday. Kept aerobic activity at 40%-42.5% of HRR for 10 minutes. Only performed 10 repetitions of strength activities. White blood cell count was low so patient received a G-CSF shot. She is experiencing some bone pain. Trainer made sure that all equipment was cleaned before client use.
4	40%-42.5% HRR 122-124 bpm RPE: 2 2 days per week 10 minutes duration	ROM 2 days per week 1 set 10 repetitions RPE: 2	Patient experiencing more nausea and fatigue than after the first treatment. Also experiencing a lot of bone pain from G-CSF shots. Workouts included a 10-minute aerobic session followed by ROM exercises and a long flexibility session.

MONTH TWO			
Week	**Aerobic**	**Strength/Endurance**	**Health Status**
1	40%-42.5% HRR 122-124 bpm RPE: 2 2 days per week 10 minutes duration	ROM, light free weights, light resistance bands, machines 2 days per week 1 set 10 repetitions RPE: 2	Patient feeling better but still experiencing some bone pain. The exercise session included 10 min of aerobic activity with ROM activities, light free weights, 10 repetitions. Because the patient does not have any imbalances in the lower body, she did perform some of her leg exercises on machines. Patient states that she feels better at the end of the session.
2	40%-42.5% HRR 122-124 bpm RPE: 2 2 days per week 10 minutes duration	ROM, light free weights, light resistance bands, machines 2 days per week 1 set 10 repetitions RPE: 2	Patient underwent third cycle of chemotherapy on Monday. She is experiencing extreme fatigue. Blood draw showed a decreased red blood cell count so patient was given Procrit. Because of extreme fatigue patient was kept at lower ranges of workout. Again ROM activities, light free weights, and the leg press were used with patient performing 1 set of 10 repetitions. Flexibility and relaxation ended the session.
3	40%-42.5% HRR 122-124 bpm RPE: 2 2 days per week 10 minutes duration	ROM 2 days per week 1 set 10 repetitions RPE: 2	Patient experiencing extreme fatigue and bone pain. Her white and red blood cell counts were low so she received a G-CSF and Procrit shot. The patient walked in the halls for 10 min, performed ROM activities, and then had a long flexibility session. All activities were performed in another room away from the other patients to decrease exposure.

(continued)

Week	Aerobic	Strength/Endurance	Health Status
4	40%-42.5% HRR 122-124 bpm RPE: 2 2 days per week 10 minutes duration	ROM, light free weights, light resistance bands 2 days per week 1 set 10 repetitions RPE: 2	Patient feeling better but still fatigued. Bone pain has lessened and white blood cell count is better, but red blood cell count is low. Received another dose of Procrit. Session again included 10 min of aerobic activity, light free weight, leg press, ROM activities, and flexibility. Patient states that she does not feel like she has lost any strength, she is just tired.

		MONTH THREE	
Week	**Aerobic**	**Strength/Endurance**	**Health Status**
1	40%-42.5% HRR 122-124 bpm RPE: 2 2 days per week 10 minutes duration	ROM, light free weights, leg machines, resistance bands 2 days per week 1 set 10 repetitions RPE: 2	Patient underwent fourth chemotherapy treatment on Monday. Patient still experiencing fatigue, but all blood counts within normal ranges. Patient still performing low range of workout and is performing all exercises with great form.
2		ROM and flexibility activities at home	Both blood counts were low and patient was experiencing severe fatigue and is starting to get a cold. Physician has advised her not to exercise until her cold is better and her blood counts return to within normal ranges. The trainer e-mailed her a home-based program for ROM and flexibility.
3	40%-42.5% HRR 122-124 bpm RPE: 2 2 days per week 10 minutes duration	ROM, light free weights, leg machines, resistance bands 2 days per week 2 sets 10 repetitions RPE: 2	Red blood count still low so patient still experiencing fatigue. Procrit was again administered. Patient does not feel as weak as last week when she had a cold. Patient performed 2 sets of the low-level workout and was comfortable with the extra workload.
4	40%-42.5% HRR 122-124 bpm RPE: 2 2 days per week 12 minutes duration	ROM, light free weights Light resistance bands 2 days per week 2 sets 10 repetitions RPE: 2	Patient received her first Taxotere treatment Monday. She feels her fatigue and bone pain have lessened. The aerobic portion of the workout was lengthened to 12 min, and the patient again performed 2 sets of the resistance and ROM exercises. She stated that she felt great when she finished her workout.

		MONTH FOUR	
Week	**Aerobic**	**Strength/Endurance**	**Health Status**
1	40%-42.5% HRR 122-124 bpm RPE: 2 2 days per week 12 min duration	ROM, free weights, leg machines, resistance bands 2 days per week 2 sets 10 repetitions RPE: 2	Patient received her second Taxotere treatment Monday. Her red blood cell count was low so Procrit was administered. She is experiencing a little more fatigue than last week but feels like she can work out at the previous week's level.
2	40%-42.5% HRR 122-124 bpm RPE: 2 2 days per week 12 min duration	ROM, free weights, leg machines, resistance bands 2 days per week 2 sets 10 repetitions RPE: 2	Patient received her third Taxotere treatment Monday. Her white blood cell count is low, but red blood cell count is within normal ranges. G-CSF was administered. She is experiencing bone pain but not as severe as before. She says her fatigue level is decreasing, but exercise session will remain at this level until her blood counts return to normal ranges.

Week	Aerobic	Strength/Endurance	Health Status
3	40%-42.5% HRR 122-124 bpm RPE: 2 2 days per week 12 min duration	ROM, free weights, leg machines, resistance bands 2 days per week 2 sets 12 repetitions RPE: 2	Patient received her fourth Taxotere treatment Monday. Her blood counts were within normal ranges. She is experiencing some fatigue, but it is lessening. She is feeling stronger. Patient performed the lower-level exercise program but performed 12 repetitions of each ROM and resistance exercise. Patient feels she has lost strength in her hands. Specific exercises were performed for this.
4	42.5%-45% HRR 124-126 bpm RPE: 2+ 2 days per week 12 min duration	ROM, free weights, resistance bands 2 days per week 2 sets 12 repetitions RPE: 2+	Patient received her fifth Taxotere treatment Monday. Her blood counts remain within normal ranges. She is still experiencing fatigue but feels like she can boost her workout a little. The intensity of the aerobic session as well as some of the resistance exercises was increased slightly. Patient was monitored closely while performing exercises in which she had to hold onto a free weight. She is still experiencing some weakness in her grip. Grip strength exercises were performed.

*Continue to emphasize whole-body workout. Start to define what weaknesses have occured because of the treatments. Begin integrating specific interventions for the weak areas in month 5.

MONTH FIVE			
Week	Aerobic	Strength/Endurance	Health Status
1	42.5%-45% HRR 124-126 bpm RPE: 2-3 2 days per week 12 min duration	ROM, free weights, leg machines, resistance bands 2 days per week 2 sets 12 repetitions RPE: 2-3	Patient received her sixth Taxotere treatment. Patient is experiencing some tightness under her right arm and weakness in the same arm. She still feels like she has lost grip strength in both hands. Patient will continue using free weights for the upper body so that any imbalances she has developed can be corrected.
2	42.5%-45% HRR 124-126 bpm RPE: 2-3 2 days per week 15 min duration	ROM, free weights, leg machines, resistance bands 2 days per week 2 sets 12 repetitions RPE: 2-3	Patient received her seventh Taxotere treatment. Patient is experiencing some fatigue but feels the lower doses of the chemotherapy have allowed her body to adjust better. The patient did 15 min of aerobic activity today without any trouble. She would like to work out 3 days a week starting next week. There is a continued emphasis on correcting the imbalances that have developed in the upper body. She is definitely stronger on the left side, so that side is maintaining strength while the right side is being restrengthened.
3	45%-47.5% HRR 126-128 bpm RPE 2-3 3 days per week 15 min duration	ROM, free weights, leg machines, resistance bands 3 days per week 2 sets 12 repetitions RPE: 2-3	Patient received her eighth Taxotere treatment. Patient is feeling fatigued but wants to try 3 days per week. The program will remain the same as last week. ROM and strength in the upper body are still being stressed. So far those are the only areas where the patient has noticed a difference physically.
4	45%-47.5% HRR 126-128 bpm RPE: 3-4 3 days per week 15 min duration	ROM, free weights, leg machines, resistance bands 3 days per week 3 sets 12 repetitions RPE: 2-3	Patient received her ninth Taxotere treatment. She tolerated the addition of another day of exercise well. She said she felt great after each workout. The imbalances in the upper body are decreasing. She has gained some strength back in her right arm. She still feels like that side is tighter than it used to be.

(continued)

Exercise Program for Jane Doe During Chemotherapy Treatment *(continued)*

MONTH SIX			
Week	**Aerobic**	**Strength/Endurance**	**Health Status**
1	47.5%-50% HRR 128-130 bpm RPE: 3 3 days per week 20 min duration	ROM, free weights, leg machines, resistance bands 3 days per week 2 sets 12 repetitions RPE: 3	Patient received her 10th Taxotere treatment. Her blood counts continue to fall within normal ranges. She feels like she is gaining more flexibility and strength in the right arm. Patient would like to start doing 20 min of aerobic activity because she was a runner before her cancer and she really enjoys the aerobic activities.
2	47.5%-50% HRR 128-130 bpm RPE: 3 3 days per week 20 min duration	ROM, free weights, resistance bands, machines 3 days per week 2 sets 12 repetitions RPE: 3	Patient received her 11th Taxotere treatment. Her blood counts continue to be good. She is enjoying the additional minutes of aerobic activity and has not displayed any adverse reactions (i.e., increased fatigue) from it. The imbalances in strength in the upper body have balanced out, so machines for the upper body have been added to her workout. Free weights are still an important part of her workout, but the machines add variety.
3	50%-52.5% HRR 130-132 bpm RPE: 3 3 days per week 20 min duration	ROM, free weights, resistance bands, machines 3 days per week 2 sets 12 repetitions RPE: 3	Patient received her 12th and last Taxotere treatment! She is thrilled. The port will be removed next week. The patient has tolerated the Taxotere well. She is still feeling some fatigue but feels the exercise has decreased that. She actually feels that in some areas she is stronger than she was before treatment. The intensity of the aerobic activity has increased slightly.
4	50%-52.5% HRR 130-132 bpm RPE: 3-4 3 days per week 20 min duration	ROM, free weights, resistance bands, machines 3 days per week 2 sets 12 repetitions RPE: 3-4	Patient had port removed on Monday, so she was advised not to do upper body exercises until the incision healed. So the session consisted of an aerobic activity, lower body work, and flexibility. The patient has value within normal ranges for all blood counts. She is still feeling that under the right arm is tighter than the left. Patient will start 7 weeks of radiation in 3 weeks. She will have a radiation treatment every weekday.

After the six months of chemotherapy and then a three-week pause in the treatment regimen, the patient began her seven weeks of radiation. The exercise reassessment was delayed until the end of her radiation treatments. The reassessment thus occurred at the end of the patient's second exercise program. The second exercise program was during radiation.

Second Program for Patient Still Undergoing Treatment

The patient's radiation treatments began during the fourth week of the second exercise intervention program. She tolerated the radiation well for the first four weeks. She did have some fatigue, but it was not nearly as severe as with the first chemotherapy protocol. Starting in the fifth week

of radiation, the patient stated that the skin under and around her right breast was breaking down. It was uncomfortable for her to wear a bra, so she performed all of the aerobic activities on bicycles to prevent any bouncing. She was not able to wear the heart rate monitor, so heart rate was obtained by palpation. Jane felt that again she had lost some strength and quite a lot of flexibility in the right arm area. Therefore in addition to her whole-body workouts, she used free weights for the upper body to return her body to strength balance between the right and left arms.

As before, the exercise specialist continued to monitor the patient prior to each session, assessing her health and fatigue status from her treatments and adjusting the workload accordingly. Progression remained slow because of the fluctuations that occurred with the treatments.

Exercise Program for Jane Doe During Treatment Following Six Months of Chemotherapy

			MONTH 7
Week	**Aerobic**	**Strength/Endurance**	**Health Status**
1	50%-52.5% HRR 130-132 bpm RPE: 3-4 3 days per week 20 min duration	ROM, free weights, resistance bands, machines 3 days per week 2 sets 12 repetitions RPE: 3-4	Feeling great, fatigue is decreasing, able to perform workouts, and feels energized following workout.
2	50%-52.5% HRR 130-132 bpm RPE: 3-4 3 days per week 20 min duration	ROM, free weights, resistance bands, machines 3 days per week 2 sets 12 repetitions RPE: 3-4	Still feeling great.
3	52.5%-55% HRR 132-134 bpm RPE: 3-5 3 days per week 20 min duration	ROM, free weights, resistance bands, machines 3 days per week 3 sets 12 repetitions RPE: 3-4	Still feeling great.
4	52.5%-55% HRR 132-134 bpm RPE: 3-5 3 days per week 20 min duration	ROM, free weights, resistance bands, machines 3 days per week 3 sets 12 repetitions RPE: 3-4	Started radiation this week. Will have radiation treatments every weekday. Experienced some fatigue toward the end of the week but was able to perform all workouts and still feel energized.
			MONTH 8
Week	**Aerobic**	**Strength/Endurance**	**Health Status**
1	45%-47.5% HRR 126-128 bpm RPE: 3 3 days per week 15 min duration	ROM, free weights, leg machines, resistance bands 3 days per week 2 sets 12 repetitions RPE: 3	Client experienced fatigue over the weekend, so the CES decreased both the volume and intensity of the cardio and strength programs. Client stated that she felt better after the workout but not totally energized.
2	45%-47.5% HRR 126-128 bpm RPE: 3 3 days per week RPE: 3 15 min duration	ROM, free weights, leg machines, resistance bands 2 days per week 2 sets 12 repetitions RPE: 3	Client stated that she wants to lift only 2 days per week because she is starting to feel stiff and sore from the radiation. CES has increased ROM exercises to hopefully prevent any loss of ROM. At the end of the week client stated that she felt fine. The exercises seem appropriate while undergoing radiation.

(continued)

Week	Aerobic	Strength/Endurance	Health Status
3	45%-47.5% HRR 126-128 bpm RPE: 3 3 days per week 15 min duration	ROM, free weights, leg machines, resistance bands 2 days per week 2 sets 12 repetitions RPE: 3	Client still fatigued, but fatigue has not increased. She still feels the stiffness and soreness but has not developed any skin irritations. Even though client has decreased amount of strength/endurance training, she does not feel like she is losing strength.
4	45%-47.5% HRR 126-128 bpm RPE: 3 3 days per week 15 min duration	ROM, free weights, leg machines, resistance bands 2 days per week 2 sets 12 repetitions RPE: 3	After 2 treatments this week the client started to notice that the skin under her right breast and armpit is becoming very tender. CES is careful the equipment or bands do not rub against the client's skin. Fatigue is still present but client feels the workouts are helping her.

MONTH 9

Week	Aerobic	Strength/Endurance	Health Status
1	45%-47.5% HRR 126-128 bpm RPE: 3 3 days per week 15 min duration	ROM, free weights, leg machines, resistance bands 2 days per week 1 set 12 repetitions RPE: 3	Client has told CES that the skin underneath her right breast and armpit is breaking down, and she is unable to wear a bra or the heart rate monitor. All aerobic activities were performed on the bicycle to decrease breast movement. Pulse was palpated. Client also requested that she do only 1 set of the resistance exercises because of the discomfort. The client did not reach her THR because she perspired too much and this irritated the breast area, so intensity was decreased to a lower level.
2	42.5%-45% HRR 124-126 bpm RPE: 2 2 days per week 15 min duration	ROM, free weights, leg machines, resistance bands 2 days per week 1 set 12 repetitions RPE: 3	Client has had an increase in skin irritation, so she has decreased the days per week of exercise for both modalities. The perspiration and irritation from clothing decrease her enjoyment and cause her stress, so until her skin heals she wants to work out only 2 days per week. CES is encouraging the client to continue with the ROM exercises to prevent a loss of ROM. Client agrees. This is the client's last week of radiation treatments.
3	42.5%-45% HRR 124-126 bpm RPE: 2 2 days per week 15 min duration	ROM, free weights, leg machines, resistance bands 2 days per week 1 set 12 repetitions RPE: 3	Skin still irritated so aerobic activity is still on the bike. Toward the end of the week the client stated that the skin was feeling better but not totally healed. She feels that she has lost strength and a lot of flexibility in the right arm. She is also experiencing some difficulty completing the 15 min of aerobic activity. Client was reassessed this week on the day of her second workout.
4	42.5%-45% HRR 124-126 bpm RPE: 2 2 days per week 10 min duration	ROM, free weights, leg machines, resistance bands 2 days per week 1 set 12 repetitions RPE: 3	Reassessment was given to client on first day of week and new goals started. Reassessment showed that the client has lost aerobic capacity; she has decreased pulmonary function and strength in shoulders and lats while she has maintained strength in all other muscle groups tested. She has a decrease in ROM in abduction, flexion, and reach in the right shoulder. She also decreased her sit and reach score, which would denote a decrease in hamstring and lower back flexibility. Skin surrounding breast has healed, and client once again can do standing aerobic activities.

Reassessment for the Exercise Prescription Following Treatment

Following radiation therapy, Jane had a reassessment (figure 7.2) so that we could obtain her fitness parameters. The results were used to write a new exercise prescription and exercise program with new goals for the patient following treatment (see pages 129-131).

Figure 7.2 Reassessment results and exercise prescription for Jane Doe following her chemotherapy and radiation treatments.

Cardiovascular Assessment Results

Protocol: Bruce treadmill test

Time completed: 5:07

Predicted $\dot{V}O_2$max: 18.52 ml · kg^{-1} · min^{-1}

Female aged 50-59

Norms for $\dot{V}O_2$max (ml · kg^{-1} · min^{-1})

Low	Fair	Good	Excellent	Superior
24	25-27	28-29	30-32	33+
18.52				

Body Composition Results

Protocol: Skinfolds (three-site) **Results:** 29%

Protocol: BIA **Results:** 33.7%

Norms for women all ages

At risk*	Above average	Average	Below average	At risk**
32%	31%-24%	23%	22%-9%	8%
	29%			

*At risk for obesity-related disorders such as heart disease

**At risk for malnutrition disorders

Pulmonary Function Results

Protocol: Forced vital capacity (FVC) **Percent of predicted:** 75

Protocol: Forced expiratory volume (FEV$_1$) **Percent of Predicted:** 71

Norms for FVC (% of predicted)

Low	Low limit of normal	Within normal limits	Excellent
<75%	75% ↔ 80%	81% ↔ 94%	95%
	75%		

MUSCULAR STRENGTH/ENDURANCE:

Test	Results	Category
Handgrip dynamometer Right (kg)	20.5	Low
Handgrip dynamometer Left (kg)	20.0	Average
Crunches (maximum)	15	Average

(continued)

(continued)

	Weight	Reps
Biceps curl L arm	10 lb	20
Biceps curl R arm	10 lb	20
Bench press	50 lb	12
Shoulder press	35 lb	8
Lat pull-down	50 lb	11
Triceps press-down	35 lb	10
Leg extension	50 lb	15
Leg curl	50 lb	20
Leg press	70 lb	18

RANGE OF MOTION (degrees):

Protocol: Goniometer

	Low	Normal	High
Right shoulder flexion (150-180)	148		
Left shoulder flexion		179	
Right shoulder extension (50-60)		58	
Left shoulder extension		57	
Right shoulder abduction (180)	166		
Left shoulder abduction		180	
Right hip flexion (100-120)		105	
Left hip flexion		110	
Right hip extension (30)	25		
Left hip extension	22		

FLEXIBILITY:

Test	Results	Category
Sit and reach (inches)	13.25	Good
Shoulder reach behind back		
Right	2	Fair
Left	3	Good

Medical/Physical concerns:

1. Mastectomy plus removal of 17 right axillary lymph nodes (4+ for cancer)
2. Chemotherapy completed 10 weeks ago
3. Completed 7 weeks of radiation therapy last week and is still experiencing minor skin irritation wearing a heart rate monitor and bra
4. Surgical scarring in right breast area and under right arm
5. Right shoulder flexion and abduction have decreased so there are marked differences betweeen sides

Relevant prescription/OTC drugs:

1. Ambien (5 mg/day) when needed
   Relevant potential side effects: drowsiness, headache, depression, dizziness, and nausea
2. Lipitor (10 mg/day)
   Relevant potential side effects: back pain, headache, abdominal pain, and constipation

EXERCISE REHABILITATION RECOMMENDATIONS

Aerobic Activity

Mode

Walking/Stationary bike.

Jane has continued to walk but not as conistently as before. We encourage herto continue walking!!!!

During rehab, use a combination of treadmill and bike per Jane's interest.

Frequency

Exercise at least 3 days per week.

Rehab sessions 3 days per week.

Try to walk on your own 1-2 days per week.

Intensity

HR should be 40%-60% HRR.

122-140 bpm.

RPE of 2-5 (moderate/strong).

Duration

10-30 min. Duration can be achieved on a combination of equipment or activities—for example, 10-min bike and 10-min walk (depends on patient interest, which may vary from day to day).

Muscle Strength/Endurance Recommendations

Start with ROM exercises especially on the right side where patient has lost some ROM.

Perform 2 sets of each exercise, 12 repetitions, 3 days per week.

RPE 3-5.

Spot ALL exercises, especially free weights.

Be mindful of pain from surgical scarring.

Make sure that the tubing and ropes do not rub on the area underneath right breast and arm.

Flexibility and Range of Motion Recommendations

Perform major muscle group stretches upon completion of exercises.

During rehab, use the wheel and rope VERY SLOWLY to increase ROM in the right shoulder region.

(continued)

(continued)

Use the wheel and rope very gently prior to weightlifting and upon completion of weightlifting.

There should be NO pain with stretching. Feel a slight pull and hold for about 10 sec.

Take circumference measurements weekly to monitor for swelling.

Perform stretches after each exercise session; hold each stretch 10-15 sec initially, breathing throughout the stretch.

TREATMENT GOALS—EXERCISE

Patient name: Jane Doe

Goals are based on the assessment and are effective if no further medical problems arise.

Initial assessment   Date: 02-29-01   **Reassessment**   Date: 11-10-01
Initial prescription   Date: 03-07-01   **Prescription**   Date: 11-17-01

	Results	Norm	6-month goals	Results	Norm	6-month goals
Body fat %	27.25	Above average	26.25	29	Above average	26.25
Endurance/Strength *Cardiovascular* (ml · kg⁻¹ · min⁻¹) Treadmill time	22.87 6:17	Low	36% ↑ 27.0 7:28	18.52 5:07	Low	27.0 7:28
Pulmonary FVC % predicted	76	Low normal	84	75	Low normal	84
FEV_1 % predicted	74		80	71		80
Handgrip (kg)	R = 25 L = 17.5	Average Low	Maintain 22	R = 20.5 L = 20.0	Low Average	25.0 22
Crunches (max)	20	Above average	31	15	Average	26
Biceps curl	R = 20 L = 17	10 lb	25 22	20 20	10 lb	25 25
Bench press	10	50 lb	15	12	50 lb	15
Shoulder press	10	35 lb	15	8	35 lb	15
Lat pull-down	14	50 lb	19	11	50 lb	19
Triceps press-down	10	35 lb	15	10	35 lb	15
Leg extension	10	50 lb	15	15	50 lb	Maintain
Leg curl	16	50 lb	21	20	50 lb	21
Leg press	15	70 lb	20	18	70 lb	20

	Results	Norm	6-month goals	Results	Norm	6-month goals
Flexibility *Sit and reach*	14.75	Excellent	Maintain	13.25	Good	14.75
Shoulder reach	R = 4 L = 3	Very good Good	Maintain 4	R = 2 L = 3	Fair Good	4 4
ROM (degrees) *Shoulder flexion*	R = 170 L = 179	Normal Normal	Maintain Maintain	R = 148 L = 179	Low Normal	170 Maintain
Shoulder extension	R = 56 L = 54	Normal Normal	Maintain Maintain	R = 58 L = 57	Normal Normal	Maintain Maintain
Shoulder abduction	R = 168 L = 180	Low Normal	180 Maintain	R = 166 L = 180	Low Normal	180 Maintain
Hip flexion	R = 101 L = 111	Normal Normal	Maintain Maintain	R = 105 L = 110	Normal Normal	Maintain Maintain
Hip extension	R = 27 L = 21	Low Low	30 30	R = 25 L = 22	Low Low	30 30

Exercise Intervention for Patients Following Treatment

The exercise intervention for patients after treatment (figure 7.3) also follows a rehabilitation model rather than a model strictly for the development of fitness. In this model patients can progress somewhat faster than during treatment because they usually do not show the severe fluctuations characteristic of the treatment phase. But progression still depends on the progress reports and the status of the patient. The first three

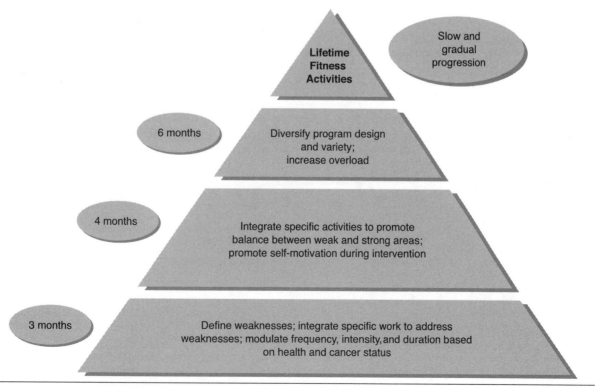

Figure 7.3 Exercise intervention following treatment. This model emphasizes a slow progression of exercises to move the patient toward precancer levels of fitness and lifetime fitness activities.

months of the exercise program for patients following treatment again focus on the whole body. The next three months continue with the whole-body work, but also incorporate exercise that is very specific to the physiological components that need to be rehabilitated in order for the patient to advance further toward an improved quality of life (e.g., range of motion work for a surgical breast cancer patient). Goal setting is appropriate and important for these patients.

Six-Month Exercise Program for a Cancer Patient Following Treatment

Jane experienced skin irritation around the right breast and under the right arm from the radiation treatments. She felt she had lost some strength and considerable range of motion in her right shoulder region, and the reassessment confirmed her observations.

Six-Month Exercise Intervention for Jane Doe Following Treatment

MONTH 10			
Week	**Aerobic**	**Strength/Endurance**	**Health Status**
1	42.5%-45% HRR 124-126 bpm RPE: 2 2 days per week 10 min duration	ROM, free weights, leg machines, resistance bands 2 days per week 1 set 12 repetitions RPE: 3	Reassessment was given to client on first day of week and new goals started. Reassessment showed that the client has lost aerobic capacity; she shows decreased pulmonary function and strength in shoulders and lats while she has maintained strength in all other muscle groups tested. She has a decrease in ROM in abduction, flexion, and reach in the right shoulder. She also decreased her sit and reach score, which would denote a decrease in hamstring and lower back flexibility. Skin surrounding breast has healed, and client once again can do standing aerobic activities.
2	42.5%-45% HRR 124-126 bpm RPE: 2 2 days per week 15 min duration	ROM, free weights, leg machines, resistance bands 2 days per week 2 sets 12 repetitions RPE: 3	Client felt that she could increase her aerobic activity again, and she was able to do it within her THR and felt great when she finished. The CES has started the client on 2 sets of resistance exercises again. She was able to tolerate this with no complaints. The CES concentrated on the shoulder region with ROM and strength exercises. The client still displays a decreased ROM on the right side. Free weights are used on the upper body to promote return to balance.
3	42.5%-45% HRR 124-126 bpm RPE: 2 2 days per week 15 min duration	ROM, free weights, leg machines, resistance bands 2 days per week 2 sets 12 repetitions RPE: 3	Client is progressing well and feeling stronger, but still feels that her ROM is decreasing on the right side. ROM exercises using the ropes and wheel are stressed.
4	45%-47.5% HRR 126-128 bpm RPE: 2-3 2 days per week 15 min duration	ROM, free weights, leg machines, resistance bands 2 days per week 2 sets 12 repetitions RPE: 3	CES has increased intensity of aerobic activity, and client loves it. Aerobically she feels she is improving.

Emphasize whole-body workouts for each session, beginning the second month following the treatment month.

MONTH 11			
Week	Aerobic	Strength/Endurance	Health Status
1	45%-47.5% HRR 126-128 bpm RPE: 2-3 2 days per week 15 min duration	ROM, free weights, leg machines, resistance bands 3 days per week 2 sets 12 repetitions RPE: 3	CES has increased strength training to 3 days per week. All upper body exercises are still done with free weights and bands to ensure that the weaker side is being equally worked. Client feels that her strength is returning on the right side. CES has noticed that she is improving in strength but the ROM on the right side is not improving. She is maintaining the ROM that she has. Again ROM exercises are stressed.
2	45%-47.5% HRR 126-128 bpm RPE: 2-3 3 days per week 15 min duration	ROM, free weights, leg machines, resistance bands 3 days per week 2 sets 12 repetitions RPE: 3	CES has increased aerobic activity to 3 days per week because client is feeling so much stronger aerobically. CES feels that balance in strength has returned to the upper body and has decided to start next week increasing strength with upper body machines along with free weights and bands. CES has started to notice slight improvements in ROM in right shoulder.
3	45%-47.5% HRR 126-128 bpm RPE: 2-3 3 days per week 15 min duration	ROM, free weights, machines, resistance bands 3 days per week 2 sets 12 repetitions RPE: 3-4	Client feels great.
4	45%-47.5% HRR 126-128 bpm RPE: 2-3 3 days per week 15 min duration	ROM, free weights, machines, resistance bands 3 days per week 2 sets 12 repetitions RPE: 3-4	Client feels great.

MONTH 12			
Week	Aerobic	Strength/Endurance	Health Status
1	42.5%-45% HRR 124-126 bpm RPE: 2 2 days per week 15 min duration	ROM, free weights, machines, resistance bands 2 days per week 2 sets 12 repetitions RPE: 2-3	Client had a busy weekend and feels like she is coming down with a cold. CES has decreased exercise intensity to prevent decreased immune function.
2	42.5%-45% HRR 124-126 bpm RPE: 2 2 days per week 15 min duration	ROM, free weights, machines, resistance bands 2 days per week 2 sets 12 repetitions RPE: 2-3	Client still has cold and is not feeling up to a strenuous work-out, but she wants to keep coming. The cold is in her head and has not traveled to her chest, so the CES felt that she could continue to exercise at the lower intensities. CES has been implementing an increase of flexibility work during this time.

(continued)

Week	Aerobic	Strength/Endurance	Health Status
3	42.5%-45% HRR 124-126 bpm RPE: 2 2 days per week 17 min duration	ROM, free weights, machines, resistance bands 3 days per week 2 sets 12 repetitions RPE: 2-3	Client still has a little head congestion but felt she could exercise aerobically a little longer. She is feeling much stronger, so CES has increased strength-training sessions to 3 days per week. Client is still increasing ROM on the right side but in small increments.
4	42.5%-45% HRR 124-126 bpm RPE: 2 2 days per week 17 min duration	ROM, free weights, machines, resistance bands 3 days per week 2 sets 12 repetitions RPE: 2-3	Congestion is gone but CES did not want to push the client so soon after her cold. Strength training is continuing along.
MONTH 13			
Week	Aerobic	Strength/Endurance	Health Status
1	42.5%-45% HRR 124-126 bpm RPE: 2 3 days per week 17 min duration	ROM, free weights, machines, resistance bands 3 days per week 2 sets 12 repetitions RPE: 2-3	Client stated that she is feeling great and would like to increase her aerobic activity to 3 times per week.
2	45%-47.5% HRR 126-128 bpm RPE: 2-3 3 days per week 17 min duration	ROM, free weights, machines, resistance bands 3 days per week 2 sets 12 repetitions RPE: 2-3	Client has started to have tingling in both hands. She has seen her physician and he thinks that she has developed neuropathy in her hands. CES is closely spotting the client when she uses free weights and resistance bands. Client is able to lift the same amount of weight that she has lifted in the past, so condition is not causing a decrease in strength. An informal grip strength test was conducted, and she has not lost any grip strength. CES will use both machines and free weights, but if a decrease in strength on either is noticed, then the CES will return to using only free weights and resistance bands.
3	45%-47.5% HRR 126-128 bpm RPE: 2-3 3 days per week 17 min duration	ROM, free weights, machines, resistance bands 3 days per week 2 sets 12 repetitions RPE: 2-3	Client still has tingling in both hands but does not have a decrease in strength on either side. Other than that, the client is feeling great and feels like her strength is increasing. The ROM on the right side has improved.
4	45%-47.5% HRR 126-128 bpm RPE: 2-3 3 days per week 20 min duration	ROM, free weights, machines, resistance bands 3 days per week 2 sets 12 repetitions RPE: 2-3	Client now experiences pain in the right arm after exercising. She states that it is not joint pain and it doesn't seem to be muscle pain but sort of a dull ache in that arm. She wonders why she doesn't have it in the other arm since she has the neuropathy in both hands. CES has referred her to her physician to have him diagnose what may be causing this pain. Client also states that the breast that was radiated is now swollen.

Week	Aerobic	Strength/Endurance	Health Status
4 (cont'd)			Again she was referred to her physician. Duration of aerobic activity has been increased, but only ROM exercises were performed on upper body until problem diagnosed.
		MONTH 14	
Week	**Aerobic**	**Strength/Endurance**	**Health Status**
1	45%-47.5% HRR 126-128 bpm RPE: 2-3 3 days per week 20 min duration	ROM, free weights, machines, resistance bands 3 days per week 2 sets 12 repetitions RPE: 2	Client has seen her physician, and he believes that the swelling in the breast area is pressing on the nerves that innervate the arm. He has prescribed an anti-inflammatory to decrease the swelling in that breast. He did not feel that she needed to stop exercising. CES monitored client's pain level during and postexercise. It was recommended that she ice the region under her arm after the exercise session. CES is stressing ROM in the shoulder and arm region. Lower intensities were used on the upper body exercises.
2	47.5%-50% HRR 128-130 bpm RPE: 3 3 days per week 20 min duration	ROM, free weights, machines, resistance bands 2 days per week 1 set 12 repetitions RPE: 2	Client still experiencing swelling in the right breast and pain in right arm after working out. CES has decreased the days per week of resistance training and also the number of sets performed because client does not enjoy the pain. She feels that the swelling is decreasing, but slowly. She still has tingling in both hands but only pain in the right arm. She still enjoys the cardiovascular workouts, and the CES has increased the flexibility portion of her intervention.
3	47.5%-50% HRR 128-130 bpm RPE: 3 3 days per week 20 min duration	ROM, free weights, machines, resistance bands 2 days per week 1 set 12 repetitions RPE: 2	Swelling in right breast has decreased and when client ices the area after working out she has very little pain in the right arm. CES will maintain her at a longer duration this week to promote more healing.
4	47.5%-50% HRR 128-130 bpm RPE: 3 3 days per week 22 min duration	ROM, free weights, machines, resistance bands 3 days per week 2 sets 12 repetitions RPE: 2	Client states that the majority of the swelling in the right breast is gone. She was able to work out the first session of this week without pain. CES has increased the number of days that she will be performing resistance exercise to 3 days per week. By the end of the week client stated that she was not having any more problems with pain but still had the tingling in her fingers.
		MONTH 15	
Week	**Aerobic**	**Strength/Endurance**	**Health Status**
1	47.5%-50% HRR 128-130 bpm RPE: 3 3 days per week 22 min duration	ROM, free weights, machines, resistance bands 3 days per week 2 sets 12 repetitions RPE: 2-3	Client is still pain free but not tingle free. She feels that it is getting better. Client states that she feels great except for the tingling in her hands.

(continued)

Six-Month Exercise Intervention for Jane Doe Following Treatment *(continued)*

Week	Aerobic	Strength/Endurance	Health Status
2	47.5%-50% HRR 128-130 bpm RPE: 3 3 days per week 25 min duration	ROM, free weights, machines, resistance bands 3 days per week 2 sets 12 repetitions RPE: 2-3	Client feels that she is stronger and more flexible than she was before her treatments. The only problem she is still exhibiting is the neuropathy in her hands, but she feels that is also improving.
3	47.5%-50% HRR 128-130 bpm RPE: 3 3 days per week 25 min duration	ROM, free weights, machines, resistance bands 3 days per week 2 sets 12 repetitions RPE: 3	Client feels great!
4	50%-52.5% HRR 130-132 bpm RPE 3-4 3 days per week 25 min duration	ROM, free weights, machines, resistance bands 3 days per week 3 sets 12 repetitions RPE: 3	Client is looking forward to her reassessment to see how far she has progressed. The ROM on the right side is approaching normal.

Final Reassessment

After the posttreatment exercise intervention, Jane again completed an assessment (figure 7.4) so that an exercise prescription could be designed for her to use at the local fitness club. As the assessment shows, Jane had decreased her percent body fat, increased her functional capacity (aerobic capacity) and time on the treadmill, and either maintained or improved on all her muscular strength except handgrip due to the neuropathy and endurance protocols. She also improved on all her flexibility and range of motion assessment protocols. At the end of her exercise intervention with us she was experiencing significantly less fatigue and had reduced her level of depression. Jane's results are consistent with those for our other patients. Patients improve their physiological status following the six-month intervention with concomitant improvements in psychological status.

INSTILLING THE LIFELONG HABIT OF EXERCISE

Exercise should be a lifelong habit. We encourage our patients to continue exercise beyond our six-month program. We have found that cancer patients do not want to leave our facility because they feel that others are not trained in working with cancer patients. For this reason we have established "satellite" sites with trained cancer exercise specialists to continue the work beyond our facility and beyond the six-month intervention. Additionally, in our workshops, we have trained many persons from other facilities. Our hope is to increase the number of facilities offering rehabilitation for cancer patients. Exercise has been shown to reduce the risk for the development of various cancers. Exercise may reduce the risk of cancer recurrence, but this issue has not been investigated.

Cardiovascular Assessment Results

Protocol: Bruce treadmill test

Time completed: 8:24

Predicted $\dot{V}O_2$max: 27.71 ml · kg^{-1} · min^{-1}

Female aged 50-59

Norms for $\dot{V}O_2$max (ml · kg^{-1} · min^{-1})

Low	Fair	Good	Excellent	Superior
24	25-27 27.71	28-29	30-32	33+

Body Composition Results

Protocol: Skinfolds (three-site) **Results:** 26%

Protocol: BIA **Results:** 31.5%

Norms for women all ages

At risk*	Above average	Average	Below average	At risk**
32%	31%-24% 26%	23%	22%-9%	8%

*At risk for obesity-related disorders such as heart disease

**At risk for malnutrition disorders

Pulmonary Function Results

Protocol: Forced vital capacity (FVC) **Percent of predicted:** 77

Protocol: Forced expiratory volume (FEV$_1$) **Percent of Predicted:** 73

Norms for FVC (% of predicted)

Low	Low limit of normal	Within normal limits	Excellent
<75%	75%-80% 77%	81%-94%	95%

MUSCULAR STRENGTH/ENDURANCE:

Test	Results	Category
Handgrip dynamometer Right (kg)	18.5	Low
Handgrip dynamometer Left (kg)	19	Average
Crunches (maximum)	19	Average

(continued)

	Weight	Reps
Biceps curl L arm	10 lb	20
Biceps curl R arm	10 lb	20
Bench press	50 lb	22
Shoulder press	35 lb	18
Lat pull-down	50 lb	15
Triceps press-down	35 lb	10
Leg extension	50 lb	22
Leg curl	50 lb	25
Leg press	70 lb	25

RANGE OF MOTION (degrees):

Protocol: Goniometer

	Low	Normal	High
Right shoulder flexion (150-180)		162	
Left shoulder flexion		179	
Right shoulder extension (50-60)		58	
Left shoulder extension		57	
Right shoulder abduction (180)	172		
Left shoulder abduction		180	
Right hip flexion (100-120)		107	
Left hip flexion		113	
Right hip extension (30)	26		
Left hip extension	23		

FLEXIBILITY:

Test	Results	Category
Sit and reach (inches)	15.25	Excellent
Shoulder reach behind back		
Right	3	Good
Left	4	Very good

Medical/Physical concerns:

1. Mastectomy plus removal of 17 right axillary lymph nodes (4+ for cancer)
2. Chemotherapy completed 10 months ago
3. Completed 7 weeks of radiation therapy 6 months ago
4. Surgical scarring in right breast area and under right arm
5. Neuropathy in both hands, decreased grip strength

Relevant prescription/OTC drugs:

Lipitor (10 mg/day)

Relevant potential side effects: back pain, headache, abdominal pain, and constipation

EXERCISE REHABILITATION RECOMMENDATIONS

Aerobic Activity

Mode

Walking/Stationary bike.

Jane has increased her walking at home up to 3 times per week. She is now walking hills. Continue this activity.

Frequency

Exercise at least 3 days per week.

Rehab sessions 3 days per week.

Try to walk on your own 3 days per week.

Intensity

HR should be 55%-65% HRR.

129-138 bpm.

RPE of 3-6 (moderate/strong).

Duratio4

25-45 min. Duration can be achieved on a combination of equipment or activities—for example, 10-min bike and 10-min walk (depends on patient interest, which may vary from day to day).

Muscle Strength/Endurance Recommendations

Perform 3 sets of each exercise, 12 repetitions, 3 days per week.

RPE 3-6.

Spot ALL exercises, especially free weights, because of the neuropathy.

Be mindful of pain from surgical scarring.

Continue ROM exercises.

Flexibility and Range of Motion Recommendations

Perform major muscle group stretches upon completion of exercises.

During rehab, use the wheel and rope VERY SLOWLY to increase ROM in the right shoulder region.

(continued)

Use the wheel and rope very gently prior to weightlifting and upon completion of weightlifting.

There should be NO pain with stretching. Feel a slight pull and hold for about 10 sec.

Take circumference measurements weekly to monitor for swelling.

Perform stretches after each exercise session; hold each stretch 15-30 sec, breathing throughout the stretch.

TREATMENT GOALS—EXERCISE

Patient name: Jane Doe

Goals are based on the assessment and are effective if no further medical problems arise.

1st reassessment   Date: 11-10-01   **2nd reassessment**   Date: 06-05-02

Prescription   Date: 11-17-01   **Prescription**   Date: 06/12/02

	Results	Norm	6-month goals	Results	Norm
Body fat %	29	Above average	26.25	26	Above average
Endurance/Strength *Cardiovascular* $(ml \cdot kg^{-1} \cdot min^{-1})$ Treadmill time	18.52 5:07	Low	27.0 7:28	27.71 8:24	Fair
Pulmonary FVC % predicted	75	Low normal	84	77	Low normal
FEV_1 % predicted	71		80	73	
Handgrip (kg)	R = 20.5 L = 20.0	Low Average	25.0 22	R = 18.5 L = 19	Low Average
Crunches (max)	15	Average	26	19	Average
Biceps curl	20 20	10 lb	25 25	20 20	10 lb
Bench press	12	50 lb	15	22	50 lb
Shoulder press	8	35 lb	15	18	35 lb
Lat pull-down	11	50 lb	19	15	50 lb
Triceps press-down	10	35 lb	15	10	35 lb
Leg extension	15	50 lb	Maintain	22	50 lb
Leg curl	20	50 lb	21	25	50 lb
Leg press	18	70 lb	20	25	70 lb

	Results	Norm	6-month goals	Results	Norm
Flexibility *Sit and reach*	13.25	Good	14.75	15.25	Excellent
Shoulder reach	R = 2 L = 3	Fair Good	4 4	3 4	Good Very good
ROM (degrees) *Shoulder flexion*	R = 148 L = 179	Low Normal	170 Maintain	R = 162 L = 179	Normal Normal
Shoulder extension	R = 58 L = 57	Normal Normal	Maintain Maintain	R = 58 L = 57	Normal Normal
Shoulder abduction	R = 166 L = 180	Low Normal	180 Maintain	R = 172 L = 180	Low Normal
Hip flexion	R = 105 L = 110	Normal Normal	Maintain Maintain	R = 107 L = 113	Normal Normal
Hip extension	R = 25 L = 22	Low Low	30 30	R = 26 L = 23	Low Low

SUMMARY

The exercise intervention is adjusted according to the treatment and health status of the patient. The cancer exercise specialist monitors the patient carefully before each exercise session and adjusts the workout accordingly. Cancer patients progress; but the progression may not be linear, and these patients may take longer to reach their goals than a healthy individual. This is not true of all cancer patients, however, so exercise programs must be individualized. Remember that an assessment should be completed initially and then reassessments should be performed following cancer therapies to help establish the next exercise intervention program.

STUDY QUESTIONS

1. What are the components of an exercise intervention?

2. Explain the exercise intervention for patients during treatment.

3. Explain the exercise intervention for patients following treatment.

4. Design an exercise intervention for a designated cancer patient.

5. What are specific risks to consider when conducting an exercise program for a cancer patient?

REFERENCES

American Cancer Society. 2002. American Cancer Society issues nutrition and activity advice for cancer survivors. Available at http://www.cancer.org. Accessed August 29, 2002.

Grantham, W.C., Patton, R.W., York, T.D., and Winick, M.L. 1998. *Health fitness management.* Champaign, IL: Human Kinetics.

Schneider, C.M., Bentz, A., and Carter, S.D. 2002a. *The influence of prescriptive exercise rehabilitation on fatigue indices.* Manuscript in preparation.

Schneider, C.M., Bentz, A., and Carter, S.D. 2002b. *Prescriptive exercise rehabilitation adaptations in cancer patients.* Manuscript in preparation.

Schneider, C.M., Dennehy, C.A., Roozeboom, M., and Carter, S.D. 2002. A model program: Exercise intervention for cancer rehabilitation. *J Integr Cancer Ther* 1(1):76–82.

Sotile, W.M. 1996. *Psychosocial interventions for cardiopulmonary patients.* Champaign, IL: Human Kinetics.

8

Establishing and Managing Cancer Rehabilitation Facilities

Despite the number of people diagnosed annually with cancer and the increasing number of those individuals who are living longer with and beyond cancer, little has been done to provide rehabilitation services to address their needs. Early detection and effective treatments are increasingly available. The new challenge is to establish quality rehabilitation services and make them available to cancer patients striving to maintain or regain strength and energy to live productive lives. The primary aim of this chapter is to outline the major issues involved in establishing or administering a cancer rehabilitation center. This information may be of most interest to readers who are proposing to undertake such a project, but it will also give cancer exercise specialists an overview of this type of program.

DEVELOPING A MISSION STATEMENT, GOALS, AND OBJECTIVES

It is important to write a clear mission statement. This statement presents the purpose of the orga-

nization and expresses the commitment to quality service that the organization makes to the community. The statement should be brief, clear, and precise; it defines the broad objectives and the purpose of the organization in relation to its products and services (Grantham, Patton, York, and Winick, 1998). For example, the mission of the Rocky Mountain Cancer Rehabilitation Institute (RMCRI) is "to advance the quality of life of cancer patients during and following treatment through prescriptive exercise rehabilitation."

Goals are targets that the facility aims to achieve. These focal points are necessary for the facility to maintain the originally intended outcomes and objectives, especially as the process unfurls. Goals of a cancer rehabilitation facility serve to guide decisions regarding operations, personnel, programs, and services. They define the facility's expectations over a given period of time, usually a year. Goals are not idealistic statements. Appropriately written goals are achievable within the context of what is known at the time they are developed and what is perceived to be possible in the future. Goals should entail outcomes that are measurable. For example, if the goal is to increase the number of referred patients seen at the facility by 10% every quarter, one need only look at the number of patients served each quarter to know whether the goal has been reached or not. These data represent the measurable outcome of the goal.

Reaching the goals in the expected timeline generally depends on the answers to several questions:

- How difficult are the goals to achieve given the resources and conditions under which the facility will operate?
- How accepting of the vision defined by the goals is the community and professional environment in which the facility operates?
- How well has the organization stayed committed to the goals?
- How valuable and appropriate to the mission are the goals?

Clearly stated objectives should define the specific strategies or actions that you will take to meet the established goals and outcomes. These strategies should be explicit and measurable. As an example, if you have the goal of increasing the number of referred patients by 10% each quarter, you might have the following objectives for reaching the goal:

1. Prepare a brochure outlining the cancer rehabilitation services and distribute the brochure to all oncologists within the community.
2. Conduct follow-up calls to all oncologists receiving brochures.
3. Advertise in the oncology newsletter given to cancer patients at the hospital.
4. Make presentations to the local service organizations in the community.
5. Offer a week of free exercise classes for cancer patients just completing their treatment.

Each objective should result in a change or an effect that brings the facility closer to the overall goals. The summative effects of all objectives should result in the intended outcomes needed to meet the goals. Figure 8.1 presents an example of a mission statement, goals, and objectives.

ADMINISTRATIVE PLANNING

During the planning phase of your program, take time to perform a thorough needs assessment and to set up a structure for your organization, as well as procedures for operation. The time spent initially will greatly add to the effectiveness of your organization.

MISSION STATEMENT, GOALS, AND OBJECTIVES	
Mission statement:	The Cancer Rehabilitation Facility is committed to improving the quality of life for cancer patients during and following treatment.
Goals:	To plan and design an integrated approach to cancer rehabilitation.
	To provide information to the community regarding the efficacy of prescriptive exercise intervention during and following cancer treatment.
	To provide patients with programs and services that meet or exceed expectations as measured by our patient assessments.
	To provide a rewarding working environment for employees that promotes low turnover and high levels of personal challenge and satisfaction.
Objectives:	To be a leader in introducing new services by investing a minimum of 10 percent of annual revenue into developing new programs and services.
	To receive recognition as a corporate leader in the community by sponsoring healthy lifestyle community events.

Figure 8.1 Sample mission statement, goals, and objectives for a cancer rehabilitation facility. Note that the objectives support the goals and the goals support the mission statement.

Needs Assessment

The first step in thinking about establishing any new endeavor is to perform a needs assessment. Cancer rehabilitation facilities should meet the unique needs of cancer patients undergoing or recovering from cancer therapies. Program offerings for patients during and following treatment should be designed and monitored by trained cancer exercise specialists. These professionals should be trained to recognize the physiological changes that occur as a result of various therapies and to monitor the tolerance level of patients for all types of exercise. For these reasons, the development and management of a comprehensive cancer rehabilitation facility should center on the three most important aspects of exercise intervention:

- Complete health and fitness assessments
- Individualized exercise prescription development
- Exercise program offerings

It is the job of those planning the facility to be sure that the following elements are in place:

- Organizational structure
- Job descriptions and personnel who perform the jobs
- Procedures
- Physical resources

All these elements are necessary to support the facility's mission of providing and implementing, on the patient level, the three major elements of exercise intervention.

Organizational Structure and Job Descriptions

For your cancer facility to be successful, you must develop an organizational structure appropriate to your needs. Within the structure, you need to establish clearly defined roles and precise job descriptions.

- **Organizational structure.** The planning process begins with determining an organizational structure that will be consistent with the facility's mission, goals, and objectives. Figure 8.2 shows an example of an organizational chart for the facility whose mission, goals, and objectives are il-

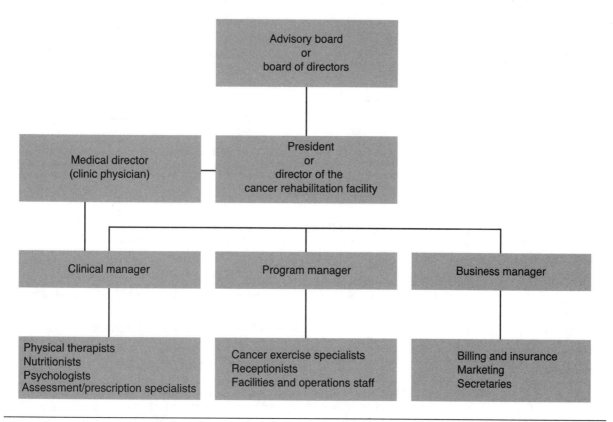

Figure 8.2 Organizational structure for cancer rehabilitation facility. This is an example of a possible structure; there are many other types. This structure has worked at RMCRI.

lustrated in figure 8.1. Notice that this plan can also be used to identify the reporting relationships among personnel. The board of directors provides a link with the community and can help your facility become established as a corporate leader. The structure also provides a designed plan that shows clearly defined responsibilities for particular personnel to help your facility operate efficiently.

• **Job descriptions.** Once you have created the structure, you need to write job descriptions defining the roles and responsibilities of the personnel in each position. This ensures that the persons hired will be well aware of their position in the organization and the responsibilities they will assume. Be sure to stress in the job description the qualifications the position requires. When you are beginning a new business that deals with a "fragile" population such as cancer patients, it becomes paramount to have trained, qualified personnel. Figure 8.3 presents job descriptions for two positions within RMCRI.

Procedures

It is important to have in place an operational plan that specifies the day-to-day functions of a cancer rehabilitation facility. The operational plan—which is the "action plan" whereby theory is translated into practice—describes in specific details the processes by which services are rendered at every level of the operation. Each person within the organization has a role to play and responsibilities to shoulder; these are highlighted in the organizational structure schema and in the job descriptions. However, what this information does

not tell you is *how* various people's responsibilities will be carried out in orchestration with others. An operational plan fills in this information.

The operational plan comprises procedures for providing each of the specific services offered at the facility, including the support services required for staff who are delivering care directly to patients.

Direct Care Procedures

The "direct care" procedures should define each aspect of the assessment, prescription development, and exercise programming and should identify the personnel responsible for the quality of the service provided. We strongly recommend that you involve personnel delivering these services in the planning process. Once you have identified the procedures for the assessment protocols, you must establish procedures for training reliable personnel who will complete the assessments. This allows for consistency in patient care even though the personnel may change. Figure 8.4 presents examples of direct care procedures.

Operational Procedures

It is important to identify the procedures that are necessary to meet the operational demands of the facility. Well-constructed and appropriate procedural information ensures that each person on staff is aware of the specific tasks involved in his job and the standards to which the tasks are to be performed. It may be best to construct procedural strategies through a team approach in which the "experts" within each aspect of the operation are encouraged to assist with the development of the

---

TWO SAMPLE JOB DESCRIPTIONS IN RMCRI

Cancer Rehabilitation Program Manager

The Program Manager is responsible for clinical rehabilitation management within RMCRI. This individual must have a master's degree in a health-related field and possess a certificate in cancer rehabilitation. The person must be able to manage personnel, direct the clinical services provided at the Institute for all cancer patients, maintain the facilities, and promote the Institute's clinical services. Exceptional leadership, communication, and management skills are required.

Cancer Exercise Specialist

The Cancer Exercise Specialist is responsible for the assessment and exercise training of all cancer patients. This individual must have a bachelor's degree in a health-related field and possess a certificate in cancer rehabilitation. The person must have a strong background in exercise physiology and have experience working within a clinical setting. Exceptional communication and personal skills are required.

Figure 8.3   Sample job descriptions, specifying responsibilities, qualifications, and skills needed for two positions within RMCRI.

plan for their area. Within cancer rehabilitation facilities, operational procedures should result in a safe environment for patients and employees, should produce patient and employee satisfaction, and should lead to financial success and sustainability. Figure 8.5 gives a list of operating procedure topics for a cancer rehabilitation facility.

SAMPLE PROCEDURES

Reliability Testing for Cancer Exercise Specialists

Cancer exercise specialists are trained to work specifically with cancer patients during and following treatment. In order to ensure accuracy and consistency during assessments, each specialist involved in the assessment of patients must undergo reliability testing. Reliability testing assures that the data collected on each patient pre- and postexercise intervention are accurate and that patterns in the techniques are consistent so that they result in reliable data. During reliability testing for assessment accuracy, the cancer exercise specialists must repeat each test protocol (on healthy subjects) until they obtain a reliability of $r = .95$ (RMCRI criteria) or better on the protocol procedures and results before working with cancer patients.

Physician Referral

New patients entering the program must have a referral prescription from their primary care physician or oncologist. A referral system ensures that the physicians and cancer exercise specialists work together to provide quality care for the patient. Once the patient's physicians refer the patient, the medical director at RMCRI and the cancer exercise specialists work together to communicate with the referring physician regarding problems, ancillary services, progress, and reassessment.

Figure 8.4 Sample direct care procedures. These paragraphs spell out procedures for reliability testing for cancer exercise specialists and patient referral at RMCRI.

OPERATING PROCEDURE TOPICS FOR CANCER REHABILITATION FACILITIES

Patient receiving and dismissal

Patient files and records

Patient scheduling

Billing and insurance

Communication

> With outside professionals
> With patients
> With employees

Facilities

Opening and closing procedures

Maintenance and quality control

Cleaning

Emergency plans

Patient assessments

Prescription development and dissemination

Exercise intervention programs

Changing and adapting to needs

Figure 8.5 Operating procedure topics for cancer rehabilitation facilities.

Safety Procedures

It is crucial that the safety precautions discussed in chapter 7 (page 116) be observed by all personnel working directly with patients. This means that you must institute regular procedures to insure that all health care workers with direct contact are meticulous in their adherence to these precautions.

It is also important for all individuals working in the cancer rehabilitation setting to be aware of and trained in the facility's emergency procedures. The plan should include each person's responsibility and the action to be taken in specific circumstances. Being able to assess the situation at hand and respond appropriately is the responsibility of all personnel in the building. Tailor a specific plan to handle anything from emergency situations to evacuation of the cancer rehabilitation facility, considering the facility's location, layout, staff coverage, location of emergency supplies and medical information files, and location of escape routes (Grantham et al., 1998).

Facilities and Equipment

The cancer rehabilitation environment requires careful planning of facilities and selection of instrumentation and equipment. There are standards that one should meet when designing a new facility or reorganizing an existing one. Each accrediting agency sets forth specific criteria for the design and maintenance of facilities. For example,

the Joint Commission on Accreditation of Healthcare Organizations accredits health care facilities so that they may provide services and qualify for reimbursements from federal agencies. The American College of Sports Medicine (1997) recommends six standards for rehabilitation facilities offering fitness-related services (figure 8.6). There are many other accrediting bodies as well, each with its own specific standards and recommendations for the facilities and the operation of those facilities. Remember, even though cancer rehabilitation is a new area, existing models of exercise intervention for individuals with an active disease or in the recovery state can help you plan, design, and develop the facility best suited for the services that you intend to provide.

In general, a comprehensive cancer rehabilitation facility should be designed to provide comfort and ease for patients. Consider each aspect of patient care and the services provided when determining the types and the amount of space you will need. For example, plan for receiving patients, conducting physical examinations and consultations, completing all health and fitness assessments, and providing sufficient space to offer various types of exercise interventions. Entrances and exits to the building should be strategically placed to accommodate emergencies and to ensure proper intake and dismissal of patients. There need to be adequate changing facilities and rest rooms, appropriately located for use by patients during testing and exercise interventions, and separate facilities for the staff. Office spaces should be planned to accommodate various per-

---

AMERICAN COLLEGE OF SPORTS MEDICINE FITNESS FACILITY STANDARDS

1. A facility should be able to respond in a timely manner to any reasonably foreseeable emergency event that threatens the health and safety of facility users. Toward this end, a facility should have an appropriate emergency plan that can be executed by qualified personnel in a timely manner.

2. A facility should offer each adult member a pre-activity screening that is appropriate to the physical activities to be performed by the member.

3. Each person who has supervisory responsibility for a physical activity program or area at the facility should have demonstrated professional competency in that physical activity program or area.

4. A facility should post appropriate signage alerting users to the risks involved in their use of those areas of a facility that present potential increased risks.

5. A facility that offers youth services or programs should provide appropriate supervision.

6. A facility should conform to all relevant laws, regulations, and published standards.

Figure 8.6 American College of Sports Medicine fitness facility standards (American College of Sports Medicine, 1997).

sonnel and the work for which they are responsible. The design should include areas that will ensure the complete confidentiality of all medical and personal records maintained on each patient. Meeting rooms should be located away from the testing and exercise areas to offer a professional environment for conferences, meetings, or training. Ample storage space should be available for all equipment and supplies and should be in a location appropriate to the ways in which the items will be used. If you intend to provide ancillary services such as massage therapy, biofeedback, or nutritional counseling, then spaces need to be so designated and need to be located for easy patient access.

Selection of equipment for use in a cancer rehabilitation facility should follow the procedures used in other rehabilitation settings. Decisions should be based on the needs of the prospective patients and the types of services you will provide. Other considerations are cost, warranties, usability, service arrangements for the equipment, and quality of the service. Since cancer rehabilitation specifically involves general physical examinations, fitness assessments, prescription development, nutritional analysis, and exercise programs, decisions regarding equipment and instrumentation purchases should be based on each of these aspects. Table 8.1 lists equipment and instrumentation for assessment and exercise training that a cancer rehabilitation facility should definitely have and that which is recommended but not required.

Table 8.1   Potential Equipment and Instrumentation

ASSESSMENT				
Body composition	Pulmonary	Cardiovascular endurance	Muscular strength and endurance	Flexibility
Skinfold calipers	Spirometer	Heart rate monitors	Handgrip dynamometer	Mats
	Metabolic cart*	Treadmill and bicycle	Hand weights	Tape measure
		Sphygmomanometer	Mats	Sit and reach box
		Stethoscope	Bench press/ Shoulder press machine	Goniometer
		Electrocardiograph*	Lat pull-down/ Seated row machine	
		Pulse oximeter*	Seated leg press machine	

TRAINING EQUIPMENT			
Cardiovascular endurance	Range of motion	Muscular strength and endurance	Monitoring equipment
Treadmill	Mats	Weight machines	Logbook
Bicycle ergometer	Ropes	Swiss balls	Anthropometric measuring tape
Recumbent leg ergometer	Range of motion wheel	Dumbbells	Heart rate monitor
Aerobic step bench	Swiss balls	Thera-Band® tubes and bands	Finger pulse oximeter

(continued)

Table 8.1 *(continued)*

TRAINING EQUIPMENT			
Cardiovascular endurance	**Range of motion**	**Muscular strength and endurance**	**Monitoring equipment**
Stationary bicycles		Xercise Bars™	Stopwatch
Spin bicycles		Rope balls	
Nordic Track		MediBalls®	
Rowing machine		Stress balls	
Cross-trainers		Weight vests	
		Free weights/Bars	
		Body bars	

*Recommended.

MANAGEMENT

The major concern of good management is to provide physical, psychological, social, and professional conditions within the organization that will optimize the services offered. In the case of a cancer rehabilitation facility, this means responding to every aspect of the operation to ensure that exercise interventions are making a positive difference in the quality of patients' lives. Management involves guiding and encouraging employees to a high standard of professional performance. It recognizes the importance of delivering excellent clinical programs in which patients consistently receive the highest standard of care. This means ensuring that the operational level of the facility is well designed and orchestrated, that the facilities and equipment are fully functional and appropriate for patient use and safety, and that communication among all levels of personnel is optimal.

Effective Management

Successful management is based on leadership and vision. A clinical manager should be knowledgeable about every aspect of the cancer rehabilitation process, from personnel and programs to finance and facilities. Every nuance of the operation is dependent on the acute understanding and responsiveness of the manager. Although this may seem to suggest that the manager needs to be all things to all people, quite the contrary is true. In fact, great managers know how to empower others to communicate information and act as needed to create an environment of trust, confidence, and pride. Remember, knowledge itself does not make an effective manager. Knowledge has to be blended with leadership, interpersonal skills, and determination.

In establishing a new cancer rehabilitation facility, new and experienced managers have a number of management techniques and training opportunities to consider from the outset. They may want to explore options for enrolling in a course or participating in leadership development workshops before assuming a new position. The Rocky Mountain Cancer Rehabilitation Institute at the University of Northern Colorado follows the theory of total quality management (Deming, 1982). The management model that most closely corresponds to the theory of total quality management is the scientific model (Grantham et al., 1998). The scientific model focuses on providing the best service in the most efficient way, which translates into high quality at low cost. At the Institute, quality management is a cooperative process that relies on the talents and skills of everyone involved in the delivery of clinical services. Quality is everyone's responsibility. Management is a continually evolving process because every interaction is unique and draws differently from the pool of knowledge, technical skills, and interpersonal relationships that everyone brings to the environment. Most importantly, personnel at the Institute recognize the value of the patients' quality of life, and through a team approach the Institute strategically and systematically provides individualized exercise interventions to improve quality of life. In this model, if the management process is effective, every aspect of clinical service will be effective.

Assessment

It is essential to develop an assessment strategy to ensure that the measures to be used will provide useful information. Data from assessments should serve as a "check and balance" between the actions taken to improve the quality of cancer rehabilitation at different levels and the perception that affected individuals have about the results. Remember that assessments should describe progress toward the outcomes sought. For example, if your clinic wants to determine how successful programs were in reducing the patients' fatigue, the assessment should be able to measure that factor. If your clinic is interested in knowing how receptive oncologists in the community are to the use of exercise intervention as a therapy, then an assessment instrument designed to measure that factor will be needed. Having unbiased ways to measure progress is an effective strategy in benchmarking progress toward the intended goals or identifying the need to develop new goals.

Training Your Personnel

The personnel working in your facility should be trained to work with cancer patients. At RMCRI we provide for training by offering one-week summer workshops and a semester curriculum course in cancer rehabilitation. The workshop includes information on cancer, cancer treatments, cancer toxicities, staging, assessment, exercise prescription development, and programming for cancer patients, and ends with a comprehensive exam. Workshop participants need a 75% success rate to obtain a cancer exercise specialist certificate. This particular workshop is available to our personnel as well as to others interested in cancer rehabilitation. Additionally, our cancer exercise specialists who complete patient assessments must successfully perform reliability testing on each assessment protocol before they can work

with cancer patients. They are trained to recognize the significance of each aspect of the assessment process. The purpose of the curriculum course is to train cancer exercise specialists in exercise programming for cancer patients, and the course involves working in the Institute with cancer patients.

ESTABLISHING COMMUNITY-BASED SUPPORT

To establish something as new as a cancer rehabilitation facility within a community, you must undertake both public relations and marketing activities to develop positive attitudes among those most in need of these services. Marketing efforts are designed to identify and satisfy consumers' wants and needs. Public relations activities build support, increase awareness, and influence opinion through effective communication strategies. Once the community recognizes the need and perceives that a cancer rehabilitation facility fills that need by providing quality services, people are far more likely to participate (Kotler, 1975).

Paramount to the success of any cancer rehabilitation effort are the recognition and support of physicians, primarily oncologists, who treat your potential clients. Establishing acceptance and developing a partnership with the oncology community

- increase patient awareness of the services provided,
- provide an opportunity for intervention both during and following treatment to minimize negative side effects, and
- improve the patient's outlook concerning recovery.

The best hope of eliciting interest in and support for this "service" is to be able to define and communicate the program's features and benefits to the population who will gain the most. To do this, you must offer the service as an answer to a problem (Wheatley, 1992). In this case, the problem is negative cancer treatment-related side effects that occur in 72% to 95% of all patients receiving treatment. You should be able to clearly make the case that a reliable, effective, and affordable exercise program will assist in managing or overcoming these effects. To demonstrate the effectiveness of your program, you will need to have pre-exercise assessments and postexercise intervention assessments that can measure the improvement in the physiological and psychological parameters that contribute to improved quality of life.

It is then important to identify the individuals within the targeted population to approach. Since oncologists will most likely be the physicians referring patients to your facility, they should be among the first to be informed of the benefits a cancer rehabilitation program can offer patients. Information about the services you provide is most effective when it is concise but detailed. Include specific physiological benefits (short- and long-term), psychological benefits, any potential risks involved, and professional supporting references. You should also describe the expertise of the staff who will be involved in delivering the services. Include brief professional biographies of the directors and managers; provide credentials of these individuals and identify their positions and associated responsibilities. It is important to clearly and succinctly describe the facility's services and programs. Include a brief description of the program (i.e., assessments performed, prescription development, nutritional analysis, etc.), detailing the purpose of each component. Lastly, the content and its presentation should leave little doubt that most cancer patients could benefit from the program and should encourage physicians to refer their patients. What is important is that you have information to present that will convince the physicians, the hospital, or the clinic to refer patients or to become part of your facility.

But physicians should not be the only group you partner with in delivering cancer rehabilitation services. Consider meeting with individuals who provide ancillary services that may also benefit patients. Because cancer rehabilitation is such a new frontier, some of the established health care providers in your community may see your endeavors as competition. In an effort to establish the best possible working relationship with other service providers in the community, it would be prudent for you to meet with them to discuss your proposed offerings to cancer patients. For example, discuss the possibility of partnering with physical therapists, massage therapists, nutritionists, and psychologists already in practice. Collaborating with established professionals in the community enables you to provide a broader menu of services. In addition, this arrangement is one that the other professionals will view as sup-

port for their businesses and that will help to establish a unique format for comprehensive care. This model puts in place a win-win working relationship within the community and provides the best services for the patient.

Since cancer affects so many people, either directly or indirectly, you should also inform the community itself of your services and the benefits you will provide to cancer patients. Many support groups, service organizations, and clinical providers can co-promote these services. The more knowledge people have about the cancer rehabilitation facility, what happens there, and how beneficial the services are, the more acceptance and support they will give it.

There are many ways to promote a newly established cancer rehabilitation facility. The most widely accepted methods for promoting health care services are through formal communication and advertising channels such as television and the print and electronic media. It is important to produce and distribute brochures. Professionally designed brochures make it easy for people to access information and are easy to distribute. When sending brochures directly to health care professionals, include an introduction letter that establishes personal contact and offers an opportunity to discuss the program. Consider including an invitation to tour the facility.

Public and media relations activities are another essential means of informing and educating others about your services. Human interest stories and information about the participants in specific aspects of a program are often well received by the population in general. Offering to speak at various types of functions (hospital staff development meetings, service organization luncheons, cancer support group meetings) is another way to introduce the community to cancer rehabilitation and its services. Public relations and marketing activities are essential during all phases of organizational growth, but particularly during the cancer rehabilitation facility's introductory years.

LEGAL ISSUES

In the field of rehabilitation, legal considerations should be a high priority. Those working in cancer rehabilitation should recognize their legal responsibilities when performing services within the cancer rehabilitation setting. The degree to which the legal expectations influence the cancer exercise specialist is in part associated with the environment where the services are being rendered. This is another reason to ensure that the management principles guiding all operations translate into high standards for every service provided, as well as consistency among those practicing and providing these services.

The cancer rehabilitation facility and one or more designated lawyers need to develop a legal services plan that outlines the areas of legal representation. The plan should include elements such as definition of the problem(s), gathering of facts, initial identification of the legal issues involved, statement of the facility's goals and expectations, list of steps necessary for success, list of uncertainties and unknowns, definition of the scope of the work, determination of required resources, forecast of schedules, definition of facility's duties, definition of lawyer's duties, range of dollar values and importance of what is at stake, evaluation of risks, determination of billing method, procedure for modifying the plan, and provisions for unknowns and changed conditions (Grantham et al., 1998).

Every country and state differs in the laws affecting responsibility during rehabilitation services. Despite these differences, fundamental legal principles apply to the cancer exercise specialist who administers health and fitness assessments, develops exercise prescriptions, and conducts exercise intervention programs. Basically, two overarching legal concepts, those of contract law and tort law, provide information regarding relationships between individuals in such settings. A valid contract should include five elements:

1. a valid offer to enter into a contract that includes the terms of the contract;

2. a valid acceptance of the offer under the terms of the contract;

3. an exchange of consideration from each party that may include money, services, actions, or promises to act or not act in a specific manner;

4. a legal purpose that does not require an illegal act or violate public policy; and

5. genuine assent between both parties that is not forced by misrepresentations or fraud (Grantham et al., 1998).

These five elements link directly to the relationships between the cancer rehabilitation facility,

the cancer exercise specialist, and the cancer patient. The basic contract should address promises by the cancer rehabilitation facility, promises by the patients, and waivers and releases. The law of torts and civil liability involves private wrongs or injuries (other than a breach of contract) for which the court awards damages. A tort is committed when a person fails to observe a duty of care or responsibility not to infringe on the patient's rights, by intentionally or carelessly causing harm or injury (Grantham et al., 1998). Examples in cancer rehabilitation are injury to the patients due to defective facilities, injury to patients due to unsafe equipment, and failure of the cancer exercise specialist to properly supervise the patient.

At the present time there are no accepted standards of practice in cancer rehabilitation. Therefore, the cancer rehabilitation facility should have a medical director (physician) who oversees all aspects of the program. However, you should review and consider standards of practice that are accepted in the general area of clinical rehabilitation. The advisory board of the cancer rehabilitation facility may want to seek legal advice about which standards to use as guidelines for clinical practice until such standards are developed and accepted in the area of exercise intervention for cancer rehabilitation. The Rocky Mountain Cancer Rehabilitation Institute requires that the patients be referred by their physicians. The Institute also utilizes a medical director as standard practice. The Institute's medical director examines each patient prior to the assessment or exercise interventions and makes recommendations as to the needs of the patient during assessment and intervention. These practices further increase assurance that the assessment and intervention will be well tolerated by and beneficial to the patient. The best way to avoid circumstances that would result in litigation is to operate under consistent practices that are designed to minimize the risk of injury or negligence and to ensure the safety and confidence of every patient.

Cancer rehabilitation facilities should operate under well-defined goals, outcomes, and objectives. These provide reasonable assurance that all clinical practices are carefully designed and based on scientific principles related to exercise intervention and cancer treatment physiology. Trained professionals who maintain current credentials in clinical exercise physiology and cancer rehabilitation should deliver all assessments and exercise interventions. No rehabilitative services should be provided to cancer patients before a written informed consent has been obtained. The informed consent verifies that the patient understands and accepts the procedures with full knowledge of the risks and benefits associated with the assessments and the exercise intervention. Obviously, consent should be obtained voluntarily from individuals of lawful age and sound mind (Herbert and Herbert, 1993). Appendix A shows a sample informed consent for cancer rehabilitation (p. 170).

REIMBURSEMENT AND FUNDING

Cancer rehabilitation is not yet a service recognized by insurance companies. However, RMCRI has been successful in obtaining reimbursement for the physician physical examination, the fitness assessment, and some rehabilitative services (physical therapy services). The cancer rehabilitation facility should obtain a provider number to receive reimbursement, or physician-owned facilities can use the physician's provider number to receive reimbursement for rehabilitation services. Primarily the facility uses the Medicare guidelines (section 2535) as the gold standard. Billing and documentation (i.e., SOAP notes: subjective, objective, assessment, and patient plan) requirements of third-party payers should be followed carefully to ensure appropriate reimbursement for services rendered. Networking with other cancer rehabilitation programs is essential to bring about awareness, knowledge, and support to obtain appropriate reimbursement. Reimbursement in other countries needs to be explored.

The Rocky Mountain Cancer Rehabilitation Institute has been successful in securing funds from local organizations such as the Susan G. Komen local Race for the Cure affiliate, which supports treatment for cancer patients. Self-pay has also been an effective way for the Institute to receive payment for services rendered. Additionally, if your facility involves research, the National Institutes of Health, the American Cancer Society, and other private and public foundations now have research money in the area of cancer survivorship. Individuals in countries outside of the United States also have possible funding sources. Explore these possibilities in your areas.

SUMMARY

Setting up or administering a cancer rehabilitation facility involves certain basic elements. Important early steps are writing the mission statement, goals, and objectives and planning for the administration of the facility. The mission statement specifies the facility's purpose; goals are targets that the facility aims to achieve, usually over a period of a year; and objectives specify the strategies or actions that will be taken to meet the goals. Administrative planning consists of assessing needs, developing the organizational structure and job descriptions, establishing procedures that describe the day-to-day functions of the facility, and planning for facilities and equipment.

Other important elements relate to management of the facility, community support, and legal and payment issues. Good management provides physical, psychological, social, and professional conditions within the organization that will optimize the services offered—in the case of a cancer rehabilitation facility, this means conditions that will result in improved quality of life for patients. Good managers are acutely aware of all aspects of the organization but also allow other professionals freedom to do their jobs. To this end managers must develop a sound assessment strategy and must ensure that personnel are properly trained. Because the idea of rehabilitation services for cancer patients is new, it is critical to inform other professionals, potential patients, and the community at large about the facility and to generate community support. Managers must understand the legal responsibilities of a rehabilitation facility and must deal with challenges pertaining to payment and reimbursement for services.

STUDY QUESTIONS

1. What is the importance of developing a mission statement, goals, and objectives for your cancer rehabilitation facility?

2. What are the primary responsibilities of a cancer exercise specialist?

3. What are some safety procedures that should be incorporated into your facility?

4. What does the American College of Sports Medicine recommend as standards for

rehabilitation facilities that offer fitness-related services?

5. Describe the instrumentation used to assess and rehabilitate specific physiological parameters for cancer patients.

6. What are the qualities of good management?

7. Why is it important to have good community support for your facility? How will you go about establishing this support?

8. What legal issues might you encounter in your facility?

REFERENCES

American College of Sports Medicine. 1997. *American College of Sports Medicine's health/fitness facility standards and guidelines* (2nd ed.), ed. J.A. Peterson and S.J. Tharrett. Champaign, IL: Human Kinetics.

Deming, W.E. 1982. *Quality, productivity, and competitive position.* Cambridge: Massachusetts Institute of Technology, Facility for Advanced Engineering Study.

Grantham, W.C., R.W. Patton, T.D. York, and M.L. Winick. 1998. *Health fitness management.* Champaign, IL: Human Kinetics.

Herbert, D.I., and W.G. Herbert. 1993. *Legal aspects of preventive and rehabilitative exercise programs* (3rd ed.). Canton, OH: Professional Reports Corporation.

Kotler, P. 1975. *Marketing for nonprofit organizations.* Englewood Cliffs, NJ: Prentice Hall.

Wheatley, M.J. 1992. *Leadership and the new science: Learning about organizations from an orderly universe.* San Francisco: Berrett-Koehler.

9

Summation: Rehabilitation of the Cancer Patient

The World Health Organization estimates that next year new cancer cases throughout the world will exceed 10 million, with 62% of these cases resulting in death (IARC Press, 2001). There are so many types of cancers that finding cures for the disease has been difficult. Rarely do you encounter a person who has not been touched by cancer. Cancer diagnosis can be devastating to the individual as well as to the entire family, and the disease of cancer has a significant impact on society as a whole. Billions of dollars are spent on prevention and treatment of the disease each year.

Additionally, cancer treatments have the potential to eliminate cancer cells, but in the process they also destroy healthy, normal cells. This alteration leads to devastating effects from the treatment itself. Toxicities often develop in many of the patient's physiological systems, leading to symptoms such as debilitating fatigue, muscle weakness, nausea, vomiting, dehydration, and psychological or emotional changes.

There is good news, however, in the fight against cancer. New screening and diagnostic technology, along with enhanced educational strategies, has led to a significant reduction in cancer mortality. Approximately 8.9 million Americans have survived cancer. As noted earlier, however, 72% to 95% (National Cancer Institute, 2002) of cancer survivors are experiencing cancer treatment-related problems.

Cancer patients are requesting participation in decision making concerning the impact of the tumor and its treatments on their functional status and quality of life. Patients fear a host of consequences, including fatigue, altered mood states, and mortality. Health care professionals have expertise in prescribing appropriate treatments and evaluating the effects of these treatments on the cancer tumor. But it has become apparent that cancer patients are requesting more from health professionals during their recovery process to help improve their quality of life. It has become essential during and following cancer treatment to develop and support programs that will help restore daily routines and promote a healthy lifestyle for those surviving cancer.

Exercise intervention to advance the quality of life of cancer patients during and following treatment has been receiving substantial support. Research on exercise rehabilitation for the cancer patient has demonstrated significant improvements in functional capacity, muscular strength and endurance, fatigue, depression, anxiety, and quality of life.

An appropriate step in cancer rehabilitation is to develop guidelines and specific programmatic components that are effective for cancer patients and survivors. At the Rocky Mountain Cancer Rehabilitation Institute (RMCRI) we have found that the essential components of a cancer rehabilitation program should include a physical examination, screening, assessment, individualized exercise prescription, a six-month exercise intervention, and reassessment. Health care professionals working with exercise and the cancer patient should be trained and certified cancer exercise specialists. Cancer patients will present for cancer rehabilitation with numerous complications due to treatment toxicities. Individuals working with these patients must have an understanding of cancer, cancer treatments, cancer treatment toxicities, and exercise as an intervention.

The previous chapters in this book have detailed the components of cancer rehabilitation that improve the quality of life for cancer patients at RMCRI. The following are some of the important points we wanted to communicate to cancer exercise specialists, current or potential:

- It is critical for cancer exercise specialists to be aware of the special considerations that need to be accounted for during cancer rehabilitation.

- Comprehensive physiological and psychological assessments form the basis for developing appropriate exercise prescriptions and exercise interventions.

- The individualized exercise prescription and intervention program are based on the type of cancer, the stage of cancer, the severity of treatment, and the time out from treatment.

- The exercise intervention is continuously monitored and is changed if the health status of the patient changes.

- The variability that is crucial to effective exercise prescriptions and exercise interven-

tions requires that all cancer exercise specialists be trained in working with cancer patients.

- The exercise prescription and exercise intervention program should be based on appropriate types, frequency, intensity, duration, and progression of exercises.

Manipulation of these parameters determines the exercise dose. It is critical to keep the exercise at a moderate level to avoid compromising the patient's health status.

Among serious diseases, cancer seems to have a mystique all its own. Even though some equally serious diseases are statistically more prevalent, people seem to fear cancer the most. Why this should be so is beyond the scope of this book to address. What we hope to have shown in this book is that the cancer rehabilitation program we describe helps cancer survivors. It helps them live with their disease, its treatment, the disease or treatment side effects, and their aftereffects. The work that patients do in their cancer rehabilitation program helps them feel better physically and psychologically. We hope that these programs proliferate and that more health professionals become involved, to continue to explore ways of encouraging cancer survivors to discover and achieve the benefits of exercise and thus enhance their quality of life.

REFERENCES

International Agency for Research on Cancer (IARC). *GLOBOCAN 2000: Cancer incidence, mortality and prevalence worldwide.* IARC CancerBase No. 5, Lyon, France: IARC Press, 2001.

National Cancer Institute. 2002. Information from PDQ for patients. Available at http://www.cancer.gov/cancerinfo/pdq /supportivecare/fatigue/patient/. Accessed September 10, 2002.

A

Forms Used at
Rocky Mountain Cancer
Rehabilitation Institute

Physical Fitness Assessment Information

Please dress in comfortable exercise clothing and tennis shoes.
(T-shirt and shorts are preferable.)

Directions to the Lab

The Exercise Physiology Laboratory is located in Gunter Hall, Room 1610, on the University of Northern Colorado campus. Gunter Hall is at the intersection of Cranford Place and 10th Avenue. If you are unfamiliar with Greeley, take Highway 34 east to 11th Avenue. Turn north (left) and continue on 11th past the University until you see Cranford Place. Turn east (right) for one block. You will be facing Gunter Hall. Please park at a meter, on the street, or in the area behind Gunter labeled "Clinic Parking." When you enter the main doors, Room 1610 will be the second door on your left on the first floor. If you parked in a clinic, permit, or meter space, please put the date and time on your parking pass and hang it on your rearview mirror.

Note: (1) Please do not participate in any type of exercise on the day of your scheduled assessment. (2) Please complete the attached paperwork and bring it with you along with your insurance card.

What to Expect

- Upon arrival, you will be asked to sign a consent form. Additional paperwork will follow.
- Your blood pressure, heart rate, height, and weight will be taken before testing begins.
- A variety of physical fitness tests will be conducted. The following information provides a brief synopsis of each of the tests you may be asked to complete.

The Components of Physical Fitness

Note: All fitness tests administered are based on the ability and any physical limitations of the individual subject. Modifications are made as needed, and the possibility exists that certain subjects may not participate in all of the tests.

Aerobic capacity, or cardiorespiratory endurance, is the ability of the heart and circulatory system to deliver oxygen from the lungs to the working tissues via the blood. Maximum oxygen uptake, or $\dot{V}O_2$max, is a measure of aerobic capacity and will be determined from heart rate and exercise time on a treadmill or stationary bicycle.

Muscular strength and endurance. Muscular strength is the ability of the muscle to generate force in one maximal effort. Muscular endurance is the ability of the muscle to generate force over a prolonged period of time. Upper and lower body strength and endurance will be measured using (a) a variety of weight machines, (b) handgrip dynamometer, and (c) maximum abdominal crunches. These tests are representative of upper and lower body muscular strength and endurance.

Flexibility is defined as the range of motion of a limb about a joint. The modified sit and reach test will be used to determine the flexibility of the hip, legs, and lower back. The goniometer, an instrument for measuring angles, will be used to measure the range of motion of the hip and shoulder joints.

Body composition is the ratio of fat tissue to lean tissue (i.e., muscle, bone, and organs) present in the body. Percent body fat will be evaluated with calipers designed to measure skinfolds at various places on the body (being dressed in T-shirt and shorts will be extremely helpful during this procedure). Excessive fat is a handicap for physical movement and can possibly lead to serious health problems, which could interfere with daily activities and job performance. Too little body fat puts an individual at risk by interfering with the formation of cell membranes, transport of necessary vitamins, and the proper function of nervous and reproductive systems.

Pulmonary function is a measurement of lung capacity. The amount of air that can be forcibly exhaled in one second indicates either healthy lungs or the possibility of an obstructive lung disorder. The subject will maximally exhale into a spirometer, and the lung capacity results will be calculated by computer.

Circumference measurements of the forearm, upper arm, lower leg, thigh, neck, and breast area will be taken. Please dress appropriately (shorts and T-shirt are preferable).

Blood and urine analyses will measure a variety of components such as immune system function, enzymes, and hormone metabolites. This portion of the assessment is optional.

The Next Step

When your physical assessment is complete, you will be asked to schedule an appointment for the following week to discuss your test results and written exercise prescription. You will also receive a computerized nutrition evaluation. Your exercise prescription will be designed to accommodate your lifestyle. If you follow the prescription, you should see increased physical fitness and zest for everyday activities.

The Definition of Physical Fitness

The President's Council on Physical Fitness and Sports defines physical fitness as "The ability to carry out daily tasks with vigor, without undue fatigue, and with ample energy to enjoy leisure-time pursuits and to meet unforeseen injuries." A satisfactory state of physical fitness should provide some protection against coronary heart disease, problems associated with obesity, various muscle and joint ailments, and physiological symptoms caused by stress and cancer.

From C. M. Schneider, C.A. Dennehy, and S.D. Carter, 2003, *Exercise and Cancer Recovery.* (Champaign, IL: Human Kinetics). Used by permission of the Rocky Mountain Cancer Rehabilitation Institute.

Cancer History

Name _____ Date _____

Address _____

City _____ State _____ Zip _____

Phone _____ Age _____ DOB _____

Type of cancer:_____ Date of diagnosis:_____

Specific location:_____(left/right breast, area of brain, etc.)

Cancer surgery: ☐ Yes   ☐ No   Type of surgery: _____

Date(s) of surgery: _____

Presenting symptom(s): _____

_____ (symptoms that led to the diagnosis of cancer: fatigue, nausea, etc.)

Postsurgery treatment (chemotherapy, radiation):

_____

Length of treatment: _____

Date of final treatment: _____

Currently undergoing chemotherapy? ☐ Yes   ☐ No

    If yes, date of most recent treatment: _____

Currently undergoing radiation? ☐ Yes   ☐ No

    If yes, date of most recent treatment: _____

Complications (infection, recurrence):_____

_____

Current medical concerns due to cancer: _____

_____

_____

Medications for cancer or cancer complications:

_____

_____

Other medications (prescribed, OTC, vitamins, herbs; if the list is extensive, ask patient to bring an inventory of meds):_____

Primary care physician at time of diagnosis: _____

Surgeon: _____

Oncologist: _____

Radiation oncologist: _____

Medical History

I. General Information

Name_____ Age_____ Date _____

Current primary care physician _____

Date of last complete physical_____

Check ALL spaces below that apply to you. (For a space you check, please include explanation and date of occurrence.)

II. Present Medical History   Explain and date:

_____Rheumatic fever/heart murmur _____

_____High blood pressure _____

_____Chest discomfort _____

_____Heart abnormalities (racing, skipping beats)_____

_____Abnormal ECG _____

_____Heart problems _____

_____Coughing up blood _____

_____Stomach or intestinal problems _____

_____Anemia _____

_____Stroke _____

_____Sleeping problems _____

_____Migraine or recurrent headaches _____

_____Dizziness or fainting spells _____

_____Leg pain after walking short distances _____

_____Back/neck pain/injuries _____

_____Foot/ankle problems _____

_____Knee/hip problems _____

_____Lymphedema _____

_____High cholesterol _____

_____Diabetes _____

_____Thyroid problems_____

_____Lung disease _____

_____Respiratory problems/asthma _____

_____Chronic or recurrent cough _____

_____Disease of arteries _____

_____Varicose veins _____

_____Increased anxiety/depression _____

_____Recurrent fatigue _____

_____Arthritis _____

_____Swollen/stiff/painful joints _____

_____Epilepsy_____

_____Vision/hearing problems _____

(continued)

Women only

_____Currently pregnant _____

_____Menstrual irregularities _____

_____Number of children _____

Last mammogram_____ Last pelvic/Pap _____

Breast self-exam: ☐ Yes   ☐ No

Operations (starting with the most recent)

1._____Date:_____

2._____Date:_____

3._____Date:_____

4._____Date:_____

Hospitalizations

(reason) _____

III. Family Medical History

_____High blood pressure   Family member(s)? _____

_____Heart attacks   Family member(s)? _____

_____Heart surgery   Family member(s)? _____

_____High cholesterol   Family member(s)? _____

_____Stroke   Family member(s)? _____

_____Diabetes   Family member(s)? _____

_____Obesity   Family member(s)? _____

_____Early death   Family member(s)? _____

_____Cancer   Type? _____ Family member? _____

                Type? _____ Family member? _____

_____Other familial illnesses (list): _____

IV. Medications
List all current medications

Medication:	Dosage:	Date started:
1._____	_____	_____
2._____	_____	_____
3._____	_____	_____
4._____	_____	_____
5._____	_____	_____

Drug allergies: _____

Data reviewed by _____
                            (signature)

Lifestyle/Activity Evaluation

Name _____ Date _____

Smoking

1. Have you ever smoked cigarettes, cigars, or a pipe? _____

2. Do you currently smoke? _____ Amount per day/week _____

3. At what age did you start smoking? _____

4. When did you quit smoking? _____

Drinking

1. During the past month, on how many days did you drink alcoholic beverages? _____

2. During the past month, what is the highest number of drinks you consumed on one occasion? _____

Sleep

1. During the past week, what was your average amount of sleep per night on the weeknights (Sunday-Thursday)? _____

2. What was your average amount of sleep on the weekend nights (Friday-Saturday)? _____

3. Would you classify your sleep as restful or restless? _____

4. Do you awake often during the night? _____

5. Do you regularly take sleep aids? _____

Exercise

1. Do you exercise on a regular basis? _____

2. What exercises do you participate in regularly? _____

3. How many days per week do you exercise regularly? _____

4. How many minutes do you spend exercising at one time? _____

5. Would you consider your exercise to be light, moderate, or vigorous? (circle one)

6. Rate your occupation:

    _____Inactive (desk job)

    _____Light work (housework, light carpentry)

    _____Heavy work (heavy carpentry, lifting)

7. What physical activities are the most enjoyable to you? _____

8. What types of facilities/equipment are available for your use? _____

9. What recreational activities (boating, camping) do you participate in? _____

10. What recreational activities do you enjoy? _____

11. Has your physical activity changed in the past year? _____

(continued)

Diet

1. Are you pleased with your current weight? _____

2. What would you like to weigh? _____ lb

3. What is the most you ever weighed as an adult? _____ lb

4. What is the least you ever weighed as an adult? _____ lb

5. What weight loss methods have you tried? _____

Daily Activity Analysis

1. Do you have difficulty performing any of the following activities?

   ____Opening jars/turning doorknobs   ____Routine yard work

   ____Carrying groceries/laundry   ____Driving

   ____Putting groceries/dishes away   ____Making a bed

   ____Removing laundry from washer/dryer   ____Lifting children

   ____Clasping any articles of clothing

   ____Other _____

2. If you answered yes to any of the above,

   a. Did the difficulty begin before or after your treatment for cancer? _____

   b. If known, explain the cause of the difficulty. _____

From C. M. Schneider, C.A. Dennehy, and S.D. Carter, 2003, *Exercise and Cancer Recovery.* (Champaign, IL: Human Kinetics). Used by permission of the Rocky Mountain Cancer Rehabilitation Institute.

Chart Contents—Initial Assessment and Reassessment

Name_____

Document	Date completed	Initials
(Right side of chart)		
Charting sheet	_____	_____
Exercise prescription	_____	_____
Data collection sheet (assessment)	_____	_____
Pulmonary printout	_____	_____
Piper Fatigue Scale	_____	_____
Beck scale	_____	_____
Informed consent	_____	_____
Cardiovascular disease risk factors	_____	_____
Lifestyle evaluation	_____	_____
Dietary analysis	_____	_____
Correspondence	_____	_____
Cancer rehab exercise logs	_____	_____
(Left side of chart)		
Problem list	_____	_____
Cancer history	_____	_____
Medical history	_____	_____
Physical exam	_____	_____
Insurance information	_____	_____

From C. M. Schneider, C.A. Dennehy, and S.D. Carter, 2003, *Exercise and Cancer Recovery.* (Champaign, IL: Human Kinetics). Used by permission of the Rocky Mountain Cancer Rehabilitation Institute.

Charting Sheet

Name _____

Date	Time	Notes	Initials

From C. M. Schneider, C.A. Dennehy, and S.D. Carter, 2003, *Exercise and Cancer Recovery*. (Champaign, IL: Human Kinetics). Used by permission of the Rocky Mountain Cancer Rehabilitation Institute.

Assessment/Reassessment Checklist

Name _____

Date _____

Technician _____

Prior to patient arrival:
1. Review chart—cancer history
2. Have HR monitor, BP cuff, stethoscope, pulse oximeter, all forms, etc., ready

Patient arrival:
1. Greet patient immediately.
2. Ask for paperwork.

    Return first page to patient.

    Check following pages for completeness: Check

    a. three days of diet diary   ☐

    b. Piper scale

      Put name and date at top   ☐

      Be sure all three pages are complete   ☐

    d. Lifestyle/Activity Evaluation   ☐

    Ask for insurance card. Reassessments, ask if insurance is the same.

    Collect copay.   ☐

Assessments
Do paperwork: Check
1. Informed Consent—you sign as witness, be sure the doctor signs   ☐
2. Medical History—write in explanations for medical conditions identified   ☐
3. Cardiovascular Risks—calculate and briefly discuss   ☐

Reassessments: Check

Complete another Medical History   ☐

Testing:

Give tests in the order outlined on data collection sheet UNLESS the doctor is waiting. If the doctor is waiting, do body composition tests first then turn the patient over to the doctor for a physical exam then immediately do the treadmill test.

When tests are complete: Check
1. Sign data collection sheet   ☐
2. Ask patients all questions on last page   ☐
3. Take final BP, HR, SpO_2 ☐
4. Schedule prescription meeting   ☐

Final procedures: Check
1. Complete ALL calculations on data collection sheet   ☐
2. Fill in pulmonary function values   ☐
3. Attach pulmonary printout to blank sheet   ☐
4. Complete Piper and Beck calculations   ☐
5. Put all pages in correct order and attach to chart   ☐
6. Diet Diary—write patient's age, sex, ht, wt   ☐
7. Place Diet Diary in nutrition tray   ☐
8. Put away HR monitor, BP cuff, stethoscope, etc.   ☐

From C. M. Schneider, C.A. Dennehy, and S.D. Carter, 2003, *Exercise and Cancer Recovery.* (Champaign, IL: Human Kinetics). Used by permission of the Rocky Mountain Cancer Rehabilitation Institute.

Informed Consent

Name_____

Date_____

The School of Kinesiology and Physical Education and the Rocky Mountain Cancer Rehabilitation Institute support the practice of protection of human subjects participating in research. The following information is provided for you to decide whether you wish to participate in the present study. You should be aware that even if you agree to participate, you are free to withdraw at any time without affecting opportunities for participation in other projects offered by this department.

This project is involved with the assessment of your cardiovascular endurance, muscular strength and endurance, range of motion, and body composition. Measuring oxygen consumption on a motor-driven treadmill will assess your cardiorespiratory capacity. Assessment of muscular strength and endurance will occur through the use of weights, dumbbells, a handgrip dynamometer, and abdominal crunches. Flexibility and range of motion are measured by an instrument called a goniometer and by the sit and reach test. The pulmonary function test requires maximum exhalation into a sterile mouthpiece. Skinfold calipers are used to measure body composition (body fat percentage). Heart rate, blood pressure, height, weight, and circumference measurements are also taken. Forms to be completed include cancer history, medical history, cardiovascular risk profile, lifestyle/activity questionnaire, and fatigue and depression scales.

Once all of the tests are completed, the results will be analyzed and an exercise prescription written. The expected benefits associated with your participation in this program include information regarding your level of physical fitness and recommended fitness and lifestyle changes necessary to improve the quality of your life.

The project will be under the direction of the Exercise Physiology Laboratory Director, but other persons will be associated with or assist with the data collection. Your participation is solicited, although strictly voluntary. The obtained data may be used in reports or publications, but your identity will not be associated with such reports.

This research should not result in physical injury; however, some soreness may occur, and some of the fitness tests can be uncomfortable. The duration of the discomfort is short.

Please give your consent with full knowledge of the nature and purpose of the procedures, the benefits that you may expect, and the discomforts and/or risks that may be encountered. We appreciate your assistance. You will be given a copy of this consent form.

_____   _____   _____
Signature of subject agreeing to participate   Signature of witness   Date
(By signing this consent you certify you are
at least 18 years of age.)

                                     _____   _____
                                     Signature of physician   Date

From C. M. Schneider, C.A. Dennehy, and S.D. Carter, 2003, *Exercise and Cancer Recovery.* (Champaign, IL: Human Kinetics). Used by permission of the Rocky Mountain Cancer Rehabilitation Institute.

Physical Exam

Doctor _____ Date _____

Patient name _____

HEENT: Pupils equal, round, reactive to light.
 θ thyromegaly, θ nucchal lymphadenopathy. Throat clear.

Exceptions:

CHEST: Cardiac—regular rate and rhythm, no murmur or gallop

 Lungs—clear without wheeze or rale

 Breast—symmetric, no masses or discharge

Exceptions:

 Last mammogram _____

ABDOMEN: Soft, nontender. No hepatosplenomegaly.
Exceptions:

GENITOURINARY: Last Pap or PSA _____

EXTREMITIES: Deep tendon reflexes 2+, no edema, symmetrical strength.

Exceptions:

Provisional diagnosis:

Vital signs: Blood pressure _____

 Heart rate _____

 Respiration _____

MD signature: _____

From C. M. Schneider, C.A. Dennehy, and S.D. Carter, 2003, *Exercise and Cancer Recovery.* (Champaign, IL: Human Kinetics). Used by permission of the Rocky Mountain Cancer Rehabilitation Institute.

Problem List

Name _____

Address _____

Phone _____

Problem list:

_____

_____

_____

_____

_____

_____

Current medications:

_____

_____

_____

_____

Allergies:

_____

_____

_____

_____

Data Collection Sheet

Name_____ Age_____ Date_____

Date of birth_____Phone_____

M/F (circle one)

Attach HR monitor immediately after patient arrives for testing.

HR monitor should not be removed until the patient completes all tests.

Start time:_____	Completion time:_____
1) BP _____mmHg	Final BP _____mmHg
2) RHR _____bpm	Final HR _____bpm
Method used _____	
Pulse oximeter reading	
3) SpO_2 _____%	Final SpO_2 _____%

4) Height _____inches   5) Weight _____pounds

                No shoes   No shoes

6) Skinfolds (body fat %)

	Abdomen	Suprailiac	Triceps	Thigh*
(female & male) 1.	_____	_____	_____	_____
*Use thigh when abdomen cannot be used. 2.	_____	_____	_____	_____
3.	_____	_____	_____	_____
Average	_____	_____	_____	_____

Sum of skinfolds _____ Witness _____

BIA _____%

7) Pulmonary function

    Be sure to record *measured* liters, not predicted liters.

    Attempt 1: FVC_____liters _____%, FEV_1 _____liters_____%

    Attempt 2: FVC_____liters _____%, FEV_1 _____liters_____%

    %var for test 2: FVC_____ If greater than 5%, repeat the test.

    Attempt 3: FVC_____liters _____%, FEV_1 _____liters_____ % IF NEEDED

8) Treadmill (Bruce protocol—terminate at 75% HRR using the Karvonen formula)

          _____([220 – age] – HRrest) (.75) + HRrest

HR termination from first assessment _____ (reassessment)

<u>New</u> termination HR = _____ (reassessment)

Today's minutes completed to reach <u>new</u> HR _____ (reassessment)

(*Note:* If patient changes from a walk to a run during this test, identify the "time" when the gait changed.)

(continued)

Bruce Protocol Worksheet

Stage	Speed	Grade	Time	BP	HR	RPE	Time	SpO$_2$
Warm-up	1.7 mph	0%	3 min					
One	1.7 mph	10%	3 min					
Two	2.5 mph	12%	3 min					
Three	3.4 mph	14%	3 min					
Four	4.2 mph	16%	3 min					
Five	5.0 mph	18%	3 min					
Recovery								

Did patient hold the handrails? ☐ Yes   ☐ No

9) Muscular endurance

A) Maximum crunches _____

B) Dynamic muscular endurance tests

Exercise	BW	×	%	=	Weight		# of reps
Biceps curl L arm	_____	×	_____	=	_____	⇒	_____
Biceps curl R arm	_____	×	_____	=	_____	⇒	_____
Bench press	_____	×	_____	=	_____	⇒	_____
Lat pull-down	_____	×	_____	=	_____	⇒	_____
Triceps press-down	_____	×	_____	=	_____	⇒	_____
Leg extension	_____	×	_____	=	_____	⇒	_____
Leg curl	_____	×	_____	=	_____	⇒	_____
Leg press	_____	×	_____	=	_____	⇒	_____
Shoulder press	_____	×	_____	=	_____	⇒	_____

% Body weight to be lifted

Exercise	Age <45 Men Women	Age 45-60 Men Women	Age 60-70 Men Women	Age >70 Men Women
Biceps curl L arm	.085 .065	.080 .061	.076 .058	.072 .055
Biceps curl R arm	.085 .065	.080 .061	.076 .058	.072 .055
Bench press	.500 .375	.470 .350	.440 .330	.410 .310

Exercise	Age <45 Men Women	Age 45-60 Men Women	Age 60-70 Men Women	Age >70 Men Women
Lat pull-down	.500 .375	.470 .350	.440 .330	.410 .310
Triceps press-down	.250 .250	.230 .230	.210 .210	.190 .190
Leg extension	.375 .375	.350 .350	.330 .330	.310 .310
Leg curl	.375 .375	.350 .350	.330 .330	.310 .310
Leg press	.750 .625	.720 .600	.690 .580	.660 .560
Shoulder press	.300 .225	.280 .210	.265 .200	.250 .185

10) Muscular strength

Handgrip dynamometer   Right hand _____, _____, _____

                                      Left hand _____, _____, _____

11) Circumference measurements (in inches)

Forearm (3 in. [7.62 cm] up from styloid process of ulna [wrist bone])   R_____   L_____

Upper arm (5 in. [12.7 cm] up from olecranon process [point of elbow])   R_____   L_____

Lower leg (5 in. [12.7 cm] up from lateral malleolus [ankle bone])   R_____   L_____

Thigh (5 in. [12.7 cm] up from superior ridge of patella)   R_____   L_____

Neck (thyroid cartilage)   _____

Subaxillary (arms abducted 90°)   _____

Across breast (arms abducted 90°)   _____

Beneath breast (bra line) (arms abducted 90°)   _____

12) Flexibility

A) Modified sit and reach _____in. _____in. _____in.   Use highest value.
B) Goniometer (degrees)

Hip setting_____   Normal Limit

        Hip flexion (standing)   R_____ L_____ 100-120

        Hip backward extension (standing)   R_____ L_____ 30

Shoulder setting_____

        Shoulder flexion (standing)   R_____ L_____ 150-180

        Shoulder backward extension (standing)   R_____ L_____ 50-60

        Shoulder abduction (standing)   R_____ L_____ 180   *(continued)*

Shoulder flexibility—shoulder reach behind back (standing)*

R_____ L_____

Norms

5 = excellent
4 = very good
3 = good
2 = fair
1 = poor

*Use the middle finger.

Technician signature _____

Is the client interested in attending cancer rehab sessions?_____

Best time of day and days/week for exercise? _____

_____

Client work hours and days/week: _____

_____

Client phone number: _____

Calculation Sheet

Body fat %_____

Women % body fat = 0.41563 (sum of 3 skinfolds; triceps, abdomen, suprailiac)
  − 0.00112 (sum of 3 skinfolds)^2 + 0.03661 (age) + 4.03653

Men % body fat = 0.39287 (sum of 3 skinfolds) − 0.00105 (sum of 3 skinfolds)^2
  + 0.15772 (age) − 5.18845

$\dot{V}O_2$max _____

$\dot{V}O_2$max calculations (Heyward, 2002, p. 61):

Male & female: Used handrails/cardiac/elderly

$\dot{V}O_2$max = 2.282 (time) + 8.545 _____ $\dot{V}O_2$max = 2.282 (_____) + 8.545 = _____

Female: Active/sedentary—no handrails

$\dot{V}O_2$max = 4.38 (time) − 3.90 _____ $\dot{V}O_2$max = 4.38 (_____) − 3.90 = _____

Men: Active/sedentary/cardiac—no handrails

$\dot{V}O_2$max = 14.76 − 1.379 (time) + 0.451 (time^2) − 0.012 (time^3)

$\dot{V}O_2$max = 14.76 − 1.379 (_____) + 0.451 (_____) − 0.012 (_____) = _____

From C. M. Schneider, C.A. Dennehy, and S.D. Carter, 2003, *Exercise and Cancer Recovery.* (Champaign, IL: Human Kinetics). Used by permission of the Rocky Mountain Cancer Rehabilitation Institute.

Assessment Results and Exercise Prescription

Name_____ Date _____

Statistics from _____

Age: _____   RHR: _____   BP: _____   Ht: _____   Wt: _____

Piper Fatigue Index score: _____   Beck Depression Inventory score: _____

Cardiovascular disease risk score: _____   (Risk _____)

Cancer: _____   Location: _____

Treatment (surgery) _____

Treatment (chemotherapy/radiation) _____

Cardiovascular Assessment Results

Protocol: Bruce treadmill test

Exercise time on treadmill: _____

Predicted 75% $\dot{V}O_2$ max: _____

Low	Fair	Good	Excellent	Superior
24	25-27	28-29	30-32	33+

Body Composition Results

Protocol: Skinfolds (three-site)   **Results:** % body fat

Norms for women all ages

At risk*	Above average	Average	Below average	At risk**
32%	31%-24%	23%	22%-9%	8%

*At risk for obesity-related disorders such as heart disease

**At risk for malnutrition disorders

Pulmonary Function Results

Protocol: Forced vital capacity (FVC) **Percent of predicted:**

Protocol: Forced expiratory volume (FEV_1) **Percent of predicted:**

Norms for FVC (% of predicted)

Low	Low limit of normal	Within normal limits	Excellent
<75%	75%-80%	81%-94%	95%

MUSCULAR STRENGTH/ENDURANCE:

Test	Results	Category
Handgrip dynamometer Right (kg)		
Handgrip dynamometer Left (kg)		
Crunches (maximum)		

	Weight	Reps
Biceps curl L arm		
Biceps curl R arm		
Bench press		
Shoulder press		

(continued)

Test	Results	Category
Lat pull-down		
Triceps press-down		
Leg extension		
Leg curl		
Leg press		

RANGE OF MOTION (degrees):

Protocol: Goniometer

	Low	Normal	High
Right shoulder flexion (150-180)			
Left shoulder flexion			
Right shoulder extension (50-60)			
Left shoulder extension			
Right shoulder abduction (180)			
Left shoulder abduction			
Right hip flexion (100-120)			
Left hip flexion			
Right hip extension (30)			
Left hip extension			

FLEXIBILITY:

Test	Results	Category
Sit and reach (inches)		
Shoulder reach behind back		
Right		
Left		

Problem List:

1.
2.
3.
4.
5.
6.
7.
8.
9.
10.

Current cancer treatment status:

Relevant prescription/OTC drugs and relevant side effects:

1.

Relevant potential side effects:

2.

Relevant potential side effects:

3.

Relevant potential side effects:

Exercise Rehabilitation Goals

Goals:

1. To strengthen the cardiovascular and pulmonary systems and reduce body fat percentage

   a. Increase treadmill time to _____ min

   b. Increase estimated 75% $\dot{V}O_2$ maximum to _____ $ml \cdot kg^{-1} \cdot min^{-1}$

   c. Improve percent predicted

   d. Decrease body fat percentage by _____

2. To increase muscular strength and endurance in all major muscle groups (upper and lower back, chest, shoulders, biceps, triceps, abdomen, and legs)

   a. Increase the number of repetitions performed on each muscular endurance exercise by _____

   b. Perform at least _____ abdominal crunches

   c. Handgrip strength _____

3. To maintain flexibility and ROM in the _____

   To improve flexibility and ROM in the_____

   a. Maintain normal values for _____

   b. Increase flexibility/ROM values

4. To reduce overall fatigue levels

   a. Follow all cancer rehabilitation guidelines

   b. Try to walk 2-3 days per week on your own

Exercise considerations:

1.

2.

3.

4.

5.

(continued)

Exercise Recommendations

Aerobic Activities

Mode

Treadmill/outdoor walking.

Weight-bearing exercise should be alternated with non-weight-bearing exercise. While weight-bearing exercise is necessary for the maintenance and improvement of muscle strength and bone density (particularly in females), it is preferable to alternate weight- and non-weight-bearing exercise in order to minimize orthopedic stress on the joints.

Frequency

Months 1-3   Months 4-6

_____ sessions/week   _____ sessions/week

Note: Initially, allow 24 hr between aerobic exercise sessions to allow for proper muscle recovery time.

Intensity

Months 1-3   Months 4-6

_____ % of HRR   _____ % of HRR

Exercise heart rate: ~   bpm   Exercise heart rate: ~   bpm

RPE of _____   RPE of _____

HRR (heart rate reserve) is the difference between estimated maximum heart rate (220 − age) and resting heart rate. The American College of Sports Medicine recommends exercise intensities of 50% to 80% of HRR (using the Karvonen formula) for the initiation of cardiovascular benefit (elevation of heart rate above resting levels is necessary for the improvement of cardio-respiratory fitness—the principle of overload). Using palpation (taking pulse) or a heart rate monitor, client should monitor heart rate throughout aerobic exercise to ensure that it remains within the prescribed beats per minute (bpm) ranges listed above. Look to increase exercise intensity at a gradual pace—progression rate will correlate with client's adaptation to consistent exercise.

Duration

Months 1-2   Months 3-4   Months 5-6

_____ min/session   _____ min/session   _____ min/session

The above numbers are guidelines that may need to be adjusted depending on the overall rate of progression (i.e., client may progress to 20 min/session sooner than _____ [date]. Initially, client may need to break down the "total" duration time into 5-6-min exercise bouts separated by a brief period of rest or flexibility training. If this is the case, keep the eventual goal of a "continuous" exercise period in mind. Maintenance of moderate exercise intensities throughout the exercise period will be very beneficial for the promotion of the use of fat as a fuel source during exercise. Increased duration time = increased caloric expenditure = possible weight loss. Total duration can be achieved on a combination of equipment or activities—for example, 5-min bike and 5-min treadmill walk (depends on client's interest, which may vary from day to day).

Muscular Strength/Endurance Activities

General guidelines:

1. Client should follow resistance training program outlined by cancer rehab specialist. For each exercise chosen, adhere to the following progression schedule:

    Month 1:

    ____set of _____repetitions to develop motor learning and proper technique

    Months 2-3:

    ____sets of ____repetitions

    Months 4-6:

    ____sets of ____repetitions

2. Strength training 2 days per week initially. Try to progress to 3 days per week after 4-5 months.

3. Rest 15-30 sec between sets.

4. RPE ____

5. Perform all exercises as "evenly" (smoothly paced) as possible (1, 2, 3 up; 1, 2, 3 down).

6. Use a "spotter" (safety partner) at all times, particularly when raising free weights above the head.

7. Maintain relatively relaxed grip throughout each lifting movement (avoid increases in blood pressure).

8. Maintain normal breathing patterns throughout each lift (never hold breath).

9. Maintain neutral posture throughout each lift.

10. Always engage in a light 5-min aerobic warm-up and a flexibility session after performing resistance training (prevention/attenuation of muscle injury and soreness).

Flexibility and Range of Motion Activities

General guidelines:

1. Client should follow flexibility routine outlined by cancer rehab specialist. Try to engage in flexibility training on a daily basis.

2. There should be no pain with stretching. Patient should feel a slight pull and hold for about 10 seconds.

3. Take circumferences measurements weekly to monitor for swelling.

4. Be sure that patient breathes throughout stretches.

5. Stretch all major muscle groups upon completion of exercise.

From C. M. Schneider, C.A. Dennehy, and S.D. Carter, 2003, *Exercise and Cancer Recovery.* (Champaign, IL: Human Kinetics). Used by permission of the Rocky Mountain Cancer Rehabilitation Institute.

Treatment Goals—Exercise

Client name_____

Goals are based on RMCRI criteria and are effective if no further medical problems arise.

Date of initial assessment:_____ Date of reassessment:_____

Date of initial prescription:_____ Date of prescription:_____

	Results	Norm	6-month goals	Results (+ represents improvement)	Norm	Next 6-month goals
Body fat %						
Cardiovascular $(ml \cdot kg^{-1} \cdot min^{-1})$ Treadmill time						
Pulmonary FVC % predicted						
FEV_1 % predicted						
Strength/Endurance *Handgrip (kg)*	R = L =			R = L =		
Abdominal crunches						
Biceps curl R	reps	lb	reps	reps	lb	reps
Biceps curl L	reps	lb	reps	reps	lb	reps
Bench press	reps	lb	reps	reps	lb	reps
Lat pull-down	reps	lb	reps	reps	lb	reps
Triceps press-down	reps	lb	reps	reps	lb	reps
Leg extension	reps	lb	reps	reps	lb	reps
Leg curl	reps	lb	reps	reps	lb	reps
Leg press	reps	lb	reps	reps	lb	reps
Shoulder press	reps	lb	reps	reps	lb	reps
Flexibility *Sit and reach*	in.				in.	
Shoulder reach	R = L =			R = L =		

182

	Results	Norm	6-month goals	Results (+ represents improvement)	Norm	Next 6-month goals
ROM (degrees) *Shoulder flexion*	R = L =			R = L =		
Shoulder extension	R = L =			R = L =		
Shoulder abduction	R = L =			R = L =		
Hip flexion	R = L =			R = L =		
Hip extension	R = L =			R = L =		

From C. M. Schneider, C.A. Dennehy, and S.D. Carter, 2003, *Exercise and Cancer Recovery.* (Champaign, IL: Human Kinetics). Used by permission of the Rocky Mountain Cancer Rehabilitation Institute.

B

Exercises Featured in the Text

BALL CRUNCH (FIT BALL CRUNCH)

1. Client places toes against the wall for balance, then crosses arms over chest to work not only the abdominals but also the sternocleidomastoids.

2. Client lies back over the ball; the more the person arches over the ball, the more difficult the exercise (a).

3. The client then slowly curls up, but not all the way up to an upright position. Client should gently exhale on the way up and inhale lying back down. Maintain low back contact with the ball as shown (b).

a

b

BALL OBLIQUES

1. Client places toes against the wall for balance, then places hands behind the neck for support but does not pull on the head while performing the exercise *(a)*.
2. Client lies back over the ball; the more the person arches over the ball, the more difficult the exercise.

3. Leading with the shoulder, the client slowly curls up and across toward the opposite knee. Client does not come up to a full upright position. Client should exhale on the way up and inhale while returning to the lying position. Maintain low back contact with the ball as shown *(b)*.

BALL BACK EXTENSION (FIT BALL BACK EXTENSION)

1. Client begins by curling over the ball on the stomach.
2. Client places the feet/legs more than shoulder-width apart against a wall. The hands/arms should be placed comfortably to either side as pictured *(a)*.

3. Client slowly performs a back extension until the ear, shoulder, and hip are in alignment. At peak extension, the client should feel and hold for 1 sec a slight sensation in the lower back, avoiding pain and back hyperextension. The client's pelvis should remain in contact with the ball. Encourage slow and controlled movements along with continual breathing *(b)*.

STANDING CALF RAISE

1. Client stands on the edge of a secure step or bench and slowly lets the heels drop below the top of the step *(a)*.

2. Client slowly rises up on toes and then returns to starting position. Client should hold on to something for support *(b)*.

3. Client can also do this one leg at a time or can hold free weights in one hand, which will increase the intensity of the exercise. Client should exhale on the way up and inhale when returning to the starting position.

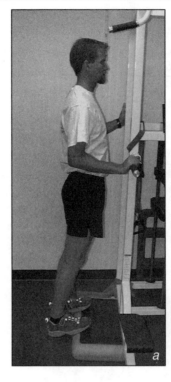

SEATED CHEST PRESS WITH SPRI BANDS (CHEST PRESS)

1. Have client sit with feet flat on floor on a bench, chair, or fit ball and grasp the handles of the bands, which are anchored around an immovable object.

2. Keeping a neutral spine throughout the exercise, the client brings the arms into a parallel position with the floor; the elbows form a 90° angle *(a)*. This parallel position is maintained during the exercise. Some clients may need trainer assistance to maintain the position.

3. Client presses the hands (palms facing down) forward and toward the midline, maintaining the arms' parallel position with the floor *(b)*.

4. Client returns to the starting position. Client should exhale when pushing forward and inhale when returning to the starting position.

BICEPS CURL WITH SPRI BANDS

1. Client stands in a staggered stance with a neutral back.
2. Client places band underneath one foot and grasps the handles in each hand, palms up *(a)*.
3. Slowly, while exhaling, client brings the hands up toward the shoulders without allowing the upper arms to move. The wrists should maintain a neutral position throughout the exercise *(b)*.
4. While inhaling, client slowly lowers the hands back to the starting position.

BENCH PRESS WITH FREE WEIGHTS

1. While holding a free weight in each hand, the client slowly lies down on the exercise bench. The trainer will have to assist by placing weights in client's hands when supine.
2. If the back has an arch in it, the client should place feet up on the bench to decrease the amount of arch. Back should remain in the beginning position throughout the exercise.
3. Client places upper arms parallel to the floor with the elbows at a 90° angle *(a)*.
4. While exhaling, client slowly pushes the free weight up toward the ceiling and in toward the midline of the body *(b)*.
5. Client then inhales and slowly lowers the weight back to the starting position.

BENCH PRESS LEARNING MOVEMENT PATTERN AND RANGE OF MOTION

1. Client lies down on workout bench with feet up on the bench if the arch in the back is great. The position the back starts in should be maintained throughout the lift.

2. Client places upper arms parallel to the floor with the elbows bent at a 90° angle *(a)*.

3. While exhaling, client pushes up toward the ceiling and in to the midline of the body *(b)*; while inhaling, the client returns to the starting position and if possible past the 90° angle at the elbow *(c)*. This will allow using a larger range of motion at this joint without any additional weight and should also cause the client to feel a slight stretch in the chest area.

BICEPS CURL WITH FREE WEIGHTS (BICEPS CURL)

1. Client places feet slightly more than shoulder-width apart with knees slightly bent, and maintains this position throughout the lifting *(a)*.

2. From the starting position, the client slowly flexes the elbow while maintaining a neutral wrist and stationary upper arms.

3. The client flexes or curls to the full range of motion, pausing at peak elbow flexion. Maintaining the spine's three curves with the ear, shoulder, and hip in alignment encourages good posture *(b)*.

LYING TRICEPS EXTENSION

1. The client assumes a supine position on a weight bench with feet flat on the floor or on the bench to maintain the low back's normal curve.

2. Place a hand weight into the client's hands *(a)*. The shoulders are maintained in 90° flexion throughout the exercise; only the elbow joint is to move *(a)*.

3. From the starting position, the client fully extends the elbow and pauses at peak elbow extension or triceps overload *(b)*.

4. The exercise is repeated as indicated. Remind the client to keep the lower back pressed into the bench and to maintain a normal breathing pattern along with proper spinal alignment.

ABDOMINAL CRUNCHES (CURL-UP CRUNCH)

1. Client begins in a supine position on a padded mat. The knees are flexed so that the feet are flat on the mat; this helps maintain the back's normal curves. The client's hands and arms are placed over the chest as shown *(a)*.

2. As the client performs trunk forward flexion, the chin is tucked toward the chest to offset neck discomfort. The client's shoulder blades will travel about 30° to 45° off the mat at peak trunk flexion *(b)*.

3. The client should feel a slight sensation in the abdominal region and hold briefly at peak trunk flexion. Also, during trunk flexion, the lower back should be "pushed" into the mat. A towel or pad may be placed under the low back if the client has trouble maintaining contact with the mat. Always encourage continuous breathing.

FRONT RAISES

1. Client begins with feet placed slightly more than shoulder-width apart and knees slightly flexed to offset lower back strain.

2. Client grasps hand weights firmly while maintaining a neutral wrist (a).

3. From the start position, the shoulder is flexed to about 90°. At this point the client should pause briefly before slowly returning to the start position (b).

4. Client returns to start position. Encourage slow, controlled movements and continuous breathing along with maintenance of good posture; only the shoulder joint moves, and the rest of the body is stationary.

INCLINE BICEPS CURL WITH FREE WEIGHTS

1. Client is positioned on an incline bench with feet placed firmly against the ground; if this is not possible, a bench or stool can be placed under the feet. Encourage the client to keep the lower back pressed firmly against the bench; tightening the abdominal region may help maintain the lower back in proper alignment throughout the exercise (a).

2. From the start position, the client performs elbow flexion to the full range of motion while maintaining a neutral wrist; only the elbow joint moves or flexes, and the rest of the body is stationary. Always encourage continuous breathing and slow, controlled movements (b).

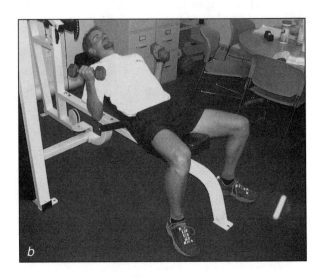

BICEPS ISOLATION/CONCENTRATION CURLS WITH FREE WEIGHTS (ONE-ARM ISOLATION BICEPS CURL)

1. Client sits on a bench with feet more than shoulder-width apart and flat against the ground.

2. Client exercises one arm at a time, thus isolating the biceps separately. In the start position, the elbow is fully extended while resting on the client's thigh (a towel can be used for extra padding if needed) *(a)*.

3. Client fully flexes the elbow while maintaining a neutral wrist and contact with the thigh *(b)*.

4. At peak elbow flexion, the client pauses briefly, then slowly returns to the start position, maintaining good posture throughout the exercise.

LATERAL RAISES

1. Client begins with a stance slightly greater than shoulder-width and knees slightly flexed.

2. While grasping the hand weights, the client maintains the elbow joints at 90° angles throughout the exercise *(a)*.

3. Client abducts the shoulder to about 90°; the elbow joint does not move and remains locked at 90° of flexion (see end position). Encourage slow, controlled movement and continuous breathing, as well as good posture, throughout the exercise *(b)*.

LEG CURL MACHINE

1. Client assumes a position so the knee joint is in line with the machine's axis of rotation (usually indicated by a red dot or other mark).

2. During the exercise, the client should keep the lower back pressed into the seat with the hands firmly grasping the handles for stability. The patient begins with legs fully extended *(a)*.

3. The patient flexes his knees and moves to peak contraction. In the end position, the cli-
ent should maintain a neutral ankle as pictured in both positions. Also during knee flexion or the overload phase, the heels are pulled toward the client's buttocks until peak range of motion is achieved, as pictured in the end position. Encourage slow, controlled movement; continuous breathing; and discourage extraneous body movement while holding peak contraction for about 4 sec *(b)*.

LEG EXTENSION MACHINE

1. Client is positioned so that the knee joint is in line with the machine's axis of rotation (usually indicated on the machine, sometimes by a red dot or other mark) *(a)*.

2. Throughout the exercise, the client should keep the lower back pressed into the seat with the hands firmly grasping the handles for stability. From the start position, the client fully extends the knee to its full range of motion, holding the full contraction for about 1 sec *(b)*.

3. Client slowly lowers the weight back to the start position. Encourage good posture; slow, controlled movement; and continuous breathing.

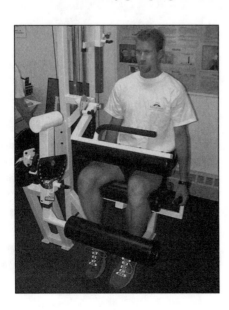

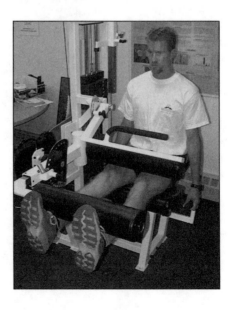

CROSSED-KNEE OBLIQUE

1. Client assumes a supine position on a mat. One knee is bent so that the foot is placed flat on the mat. The other foot is positioned on the opposite, bent knee as shown *(a)*.

2. Client places the hands gently behind the head as shown. Client should avoid pulling on the neck forcefully during the exercise; in other words, the hands should rest lightly behind the head throughout the contraction or "up" phase *(a)*.

3. To isolate the oblique muscle, the client leads with one shoulder toward the opposite leg, which is bent such that that foot rests on the other knee (as shown). Remind the client to keep the lower back placed on the mat, to breathe continuously, and to move in a slow and controlled manner *(b)*.

HEEL TAPS FOR ABDOMINALS

1. Client assumes a supine position on a mat.

2. Hips are flexed and held at about 90° in the start position *(a)*.

3. Maintaining good posture, continuous breathing, and slow and controlled movement, the client alternates between the left and right legs, performing either a bent-knee *(b)* or a straight-leg heel tap *(c)*, and then returns to the start position. Remind the client to keep the low back closely pressed against the mat; if this is not possible, a towel or soft pad may be placed under the low back.

Note: The straight-leg tap is more advanced than the bent-knee tap.

BALL PELVIC MOTIONS, ANTERIOR TO POSTERIOR (FIT BALL PELVIC MOTIONS)

1. Client takes a seated position on a Thera-Band ball or similar device.

2. Feet are placed in a firm, flat position while the arms rest comfortably at the client's side.

3. In the start position the pelvis is neutral (hips are positioned under shoulders) *(a)*.

4. To perform the exercise, the client moves the iliac crest in a posterior direction, thus gently overloading the rectus abdominis. Encourage client to maintain good posture and to use continuous breathing along with slow, controlled movements *(b)*.

BALL PELVIC MOTIONS, SIDE TO SIDE (FIT BALL PELVIC MOTIONS)

1. Client assumes a seated position on a Thera-Band ball or similar device.

2. Feet are placed in a firm, flat position while the arms rest comfortably at the client's side.

3. In the start position the pelvis is neutral (hips are positioned under shoulders) *(a)*.

4. Client alternates repetitions by moving the hip right, then left, and so on *(b)*. Remind client to maintain good posture, continuous breathing, and slow, controlled movement.

FRONT-SIDE-FRONT ARM RAISES

1. Client begins with a stance slightly greater than shoulder-width and knees slightly flexed, with arms perpendicular to the floor *(a)*.

2. Client flexes the shoulder to about 90°—or places arms parallel to the floor—and pauses *(b)*.

3. Client horizontally abducts the shoulder 90° and pauses *(c)*.

4. Client horizontally adducts the shoulders 90° *(b)*.

5. Client extends shoulder so that the arms are once again perpendicular to the floor *(a)*. Encourage client to maintain good posture, to breathe continuously, and to use slow, controlled movement.

ROPES FLEXION

1. Client stands in about a shoulder-width stance with feet offset, one slightly in front of the other.
2. Client reaches to grab the rope as shown *(a)*.
3. The exercise, range of motion movement, is performed when the overhead arm pulls down on the rope, causing the other arm/shoulder to passively flex while the other shoulder actively extends *(b)*.
4. Client alternates arms as indicated. Instruct client to breathe continuously, maintain good posture, and use slow, controlled movement throughout the exercise.

ROPES ABDUCTION

1. Client stands in about a shoulder-width stance with feet offset, one slightly in front of the other.
2. Client reaches to grab the rope as shown *(a)*.
3. The exercise, range of motion movement, is performed when the overhead arm pulls down on the rope, causing the other arm/shoulder to passively abduct while the first shoulder actively adducts *(b)*.
4. Client alternates arms as indicated. Instruct client to breathe continuously, maintain good posture, and use slow, controlled movement throughout the exercise.

LAT ROW (DUMBBELL LAT ROWS; ONE-ARM LAT ROW)

1. Initially client positions feet so that one foot is in front of the other. The back leg remains mostly straight, and the front leg is slightly bent.

2. For stability, the client reaches with one hand for a stationary object about waist high. Hip flexion is expected, and the client should be reminded to keep the abdominal muscle tight and the three natural curves in the back in alignment.

3. The other hand then grasps a hand weight; the arm begins perpendicular to the floor *(a)*.

4. Client extends the shoulder and flexes the elbow simultaneously about 90° each. At peak shoulder extension, the hand weight should be about even with the pectoralis major or chest muscle *(b)*. Throughout the exercise, the client should maintain proper posture (ear, shoulder, hip in alignment), continuous breathing, and slow, controlled movement.

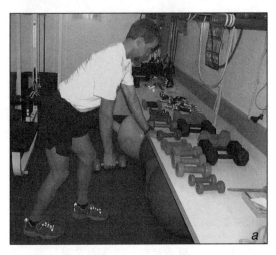

ONE-ARM LOW ROW WITH SPRI BANDS (ONE-ARM LOW ROW)

1. Client begins by sitting on a bench with good posture.

2. The knees are bent at about 90° with the feet flat on the floor as shown.

3. Next, tubing is placed around a stationary object, and the client assumes a position such that there is tension on the tubing while the client remains in the starting posture *(a)*.

4. Alternating arms, the client pulls in a horizontal and posterior direction with the hand, causing the elbow to flex and the shoulder to extend. The end position is the position in which the hand is about even with the pectoralis major muscle *(b)*.

5. The client slowly returns to the starting position and repeats the movement with the other arm. Encourage slow, controlled movement and continuous breathing.

SEATED CALF RAISE

1. Client sits on a bench using good posture (ear, shoulder, and hip in alignment). Feet are placed on a raised object about shoulder-width apart.

2. In the start position, the knees are bent at about 90° and the balls of the feet rest on the raised object. For added overload, a hand weight can be held across the top of the distal thigh (a).

3. To perform the exercise, the client rises up on the toes, performing plantar flexion in a full range of motion (b). The client should feel a mild sensation in the posterior gastrocnemius muscle. The client should use continuous breathing, good posture, and slow, controlled movement.

BACK-SUPPORTED MILITARY PRESS (MILITARY PRESS; ONE-ARM MILITARY PRESS)

1. Client begins exercise by sitting with the back supported.

2. The feet are placed firmly and flat on the ground with the knees bent at about 90°. The client maintains good posture, and the back should remain firm against the back support.

3. The client holds hand weights evenly in the two hands so that the weights start out at about shoulder height, as shown (a).

4. Alternating arms, client performs an overhead press so that the upper arm fully abducts as the elbow fully extends (b). Encourage continuous breathing, good posture, and slow, controlled movement.

PRONE OPPOSITE ARM AND LEG RAISE (SUPERMAN'S)

1. Client assumes a full prone position on a mat or other comfortable surface with the hands and feet positioned as shown (a).

2. This exercise is for back stabilization; thus progression should be gradual.

3. Initially, the client might perform only arm raises, one at a time, or leg raises, one at a time. A more advanced technique involves the end position shown (b). The client slowly raises one arm and the opposite leg off the ground simultaneously (shown in the end position (b)—right leg raised and left arm raised). The client should not feel any pain but should feel a mild sensation in the lower and upper back. If pain occurs in the lower back, discontinue this exercise until the client recovers.

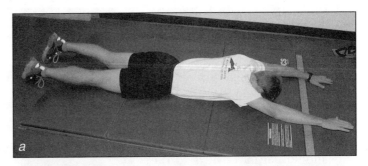

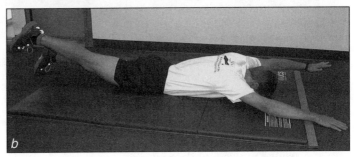

4. You can instruct the client either to repeat the same alternating combination for a number of reps or to switch from a right arm/left leg combination to a left arm/right leg combination every other repetition.

TRICEPS KICKBACK

1. Client stands with one foot in front of the other, with the back leg mostly straight and the front knee slightly bent.

2. For support, the client flexes the trunk and balances with one hand on a sturdy object, as shown (a).

3. In the start position, the client flexes the elbow to 90° and extends the shoulder so that the elbow is about even with or slightly above the side (a).

4. During the exercise, the elbow remains stationary against the side as the elbow joint is fully extended and briefly held (b) before returning to the start position. Strongly encourage continuous breathing, good posture, and slow, controlled movement.

WALL BALL SQUATS

1. Client stands with ball in the small of back (low back) pressed against the wall. Feet are shoulder-width apart and firmly flat against the ground *(a)*.

2. From the standing position, the client squats downward so that the hips and knees flex no more than 90°. Note that the knees should *not* move beyond the balls of the feet and should instead remain near or over the malleoli (anklebones) *(b)*.

3. To finish the exercise, the client slowly stands back up, maintaining a slight knee bend once back in the start/end position *(a)*. Continuous breathing, good posture, and slow, controlled movement throughout the exercise are important.

WHEEL

1. Client positions the shoulder in line with the wheel's axis of rotation.

2. Client clasps the wheel handle and flexes the shoulder as far as possible *(a)*.

3. Client then moves the wheel out of flexion and into extension of the shoulder *(b, c)*.

4. Client remains in an upright position throughout the exercise, with shoulder remaining in alignment with the wheel's axis. The client should use continuous breathing, slow and controlled movement, and good posture. This exercise should not be performed if it causes pain or discomfort.

 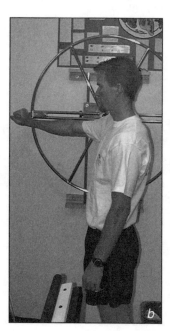 

STANDING TRICEPS EXTENSION WITH TUBING

1. Client stands with feet shoulder-width apart, in a staggered stance, and knees slightly bent *(a)*.

2. The tubing is placed over a sturdy bar so that the handles on the tubing can easily be grasped with the elbows held loosely at the side.

3. The client grasps the bands with a palm-down grip.

4. Client pushes down on the handle until the elbow is in full extension *(b)*, then releases the band back up to form a 90º angle at the elbow. Client should maintain the upper arm in the start position, with a neutral back and a neutral wrist. Good posture is realized when the client maintains the ear, shoulder, and hip in alignment. Only the elbow joint moves. The upper arm and shoulder joint remain stationary throughout the exercise. Encourage continuous breathing and slow, controlled movement.

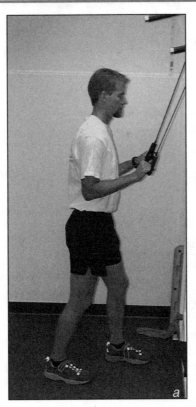
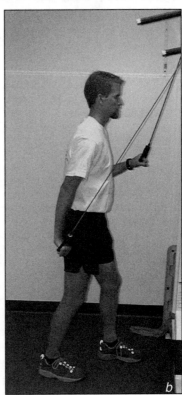

Glossary

acute—Referring to an immediate response or immediate effects.

adaptations principle—Principle stating that physiological changes occur as a result of training.

adenosine triphosphate or ATP—A complex chemical compound that is the basic source of energy.

aerobic metabolism—Production of energy through the utilization of inspired oxygen.

alkaloids—A group of anticancer drugs that prevent cell duplication by interrupting the formation of chromosome spindles.

alkylating agents—A group of anticancer drugs that combine with the DNA in a cancer cell to prevent cell division.

allogenic—Referring to tissue obtained from another individual.

aminopeptidases—Digestive enzymes involved in the catabolism of proteins.

anaerobic metabolism—The provision of energy without utilization of inspired oxygen.

anaplasia—A condition in which tissue in an organ is undifferentiated and has no resemblance to the normal tissue of the organ.

anemia—Low levels of red blood cells with concomitant low concentrations of hemoglobin.

antiangiogenic—Referring to prevention of new blood vessel formation.

antibody—A protein (gamma globulin) made by the body in response to a specific foreign protein or antigen. The antigen results from a cancer or some other source. If the foreign substance attacks again, the white blood cells are able to recognize it and reproduce the specific antibody to fight it.

antigen—A substance that causes activation of the immune system.

antimetabolites—A group of anticancer drugs that resemble normal vitamins or building blocks of metabolism. The tumor cells "think" they are getting the real vitamin or building block and thus lack sufficient nutrients to grow or multiply.

antineoplastic—Referring to a process that interferes with or prevents the growth and development of malignant cells.

antitumor antibiotics—A group of anticancer drugs that insert into strands of DNA, breaking the chromosomes or inhibiting the synthesis of ribonucleic acid (RNA).

apoptosis—Digestion by phagocytes of membrane-bound cell fragments from destroyed or disintegrated cells.

arterioles—Minute arteries that lead into the capillaries.

aspiration biopsy—Removal of tissue, usually with a needle, from a specific area of the body.

ATP-PCr system—An anaerobic metabolic system in which phosphocreatine releases energy when creatine and phosphate split.

autologous—Referring to tissue obtained from the patient.

benign tumor—A tumor that is not progressive or recurrent.

biological response modifiers—Agents that alter the immunological host tissue response when a foreign invader attacks.

blood pressure—The pressure exerted by the blood on the wall of any vessel.

B lymphocytes—Lymphocytes arising in the bone marrow that produce humoral antibodies.

carcinogen—Any substance or agent that produces or incites cancer.

carcinogenesis—The process whereby a normal cell becomes cancerous.

carcinomas—A form of cancer that develops in the epethelial tissue lining organs such as uterus, lung, and breast.

cardiac dysrhythmias—Abnormal, disordered, or disturbed cardiac rhythms.

cardiac output—The amount of blood discharged from the left or right ventricle per minute.

cell cycle—The process involved with the division of cells. The cell cycle includes mitosis, two gap phases, and a synthesis phase.

cellular immune responses—Responses that defend the body against foreign invaders attacking within the cell.

cholestasis—Arrest of bile excretions.

chromosome—The fundmental strands of genetic material (DNA) that carry our genes.

chronic—Referring to late responses or effects.

cytoplasm—The protoplasm of a cell, outside the nucleus.

cytotoxic—Referring to a process or agent that destroys cells.

deoxyribonucleic acid (DNA)—A nucleic acid present in chromosomes of the nuclei of cells that is the chemical basis of heredity and the carrier of genetic information for all organisms.

differentiation—How closely cancer cells resemble normal cells from the same organ or tissue.

disaccharidases—Digestive enzymes involved in the catabolism of carbohydrates.

duration—The amount of time work is performed during each exercise session.

dyspnea—Hunger for air resulting in labored or difficult breathing.

endothelial cells—Flat cells that line the blood and lymphatic vessels, the heart, and various other body cavities.

epithelial cells—Cells lying on the basement membrane that form the epidermis of the skin and the surface layer of mucous and serous membranes.

excisional biopsy—Surgical removal of an entire suspected tumor along with minimal surrounding normal tissue.

external radiotherapy—A therapy in which beams of radiation from a source outside the body are aimed at the cancerous area.

fibroblast growth factor—A growth factor that can cause fibroblasts to multiply and grow.

fibrosis—Abnormal formation of fibrous tissue (scarring).

fistula—An abnormal tubelike passage.

frequency—The number of exercise sessions performed per week.

gap phases—Phases within the cell cycle in which increases in protein synthesis are occurring in preparation for cell division.

genome—The complete set of chromosomes contributed by one of the male-female pair.

glycolytic system—An anaerobic metabolic system involving the release of energy through the degradation of glucose.

grading—A way of describing tumors by their appearance under a microscope.

granulocytopenia—Deficiency of granulated white blood cells.

heart rate—The number of times the heart contracts and relaxes per minute.

hematopoietic tissues—Tissues that produce blood cells, for example bone marrow.

hemolytic—Pertains to the breaking down of red blood cells.

histological examination—Examination of the appearance of tissues, such as cancers, under a microscope.

humoral immune responses—Responses from substances within the body fluids produced by the immune system that defend the body against foreign invaders.

immune effector cells—Cells within the immune system that serve many functions, one of which is to increase tumor recognition.

immunoaugmenting agents—Agents involved in the disruption of viral replication.

immunoregulating agents—Agents that activate B cells and T cells and enhance the inflammatory response.

immunorestorative agents—Agents that restore immune function.

incisional biopsy—A procedure in which a small wedge of tissue is obtained from a larger mass for diagnosis.

individuality principle—Principle stating that individual responses to training are variable.

initiation—The first stage in carcinogenesis, in which a carcinogen attacks a normal cell.

intensity—The degree of difficulty of work being performed.

internal radiotherapy—Radiation administered near the tumor within the body through radioactive isotopes.

interphase—The phase of the cell cycle in which cell division is not occurring but normal cell growth and metabolism are occurring.

intraperitoneal—Referring to administration directly into the abdominal cavity.

intravenous—Referring to administration into a vein.

leukemia—A chronic or acute disease characterized by unrestrained growth of leukocytes and their precursors in the tissue.

leukopenia—Low levels of white blood cells.

lymphatics—Pertains to the lymphs, lymph vessels, and lymph nodes.

lymphedema—Swelling, usually of an arm or leg, caused by obstructed lymphatic vessels.

lymphocytopenia—Low levels of white blood cells produced in the lymph system.

lymphomas—Pertains to the growth of new tissue in the lymphatic system.

macrophages—Cells of the reticuloendothelial system having the ability to ingest particulate substances.

malignant tumor—A tumor that is cancerous and tending to threaten death.

maximal oxygen consumption—The maximal rate at which oxygen can be utilized by active cells during high-intensity exercise.

melanomas—Malignant, pigmented moles or tumors.

metastasis—The manifestation of a malignancy in a secondary growth in a new location arising from the primary growth.

minute ventilation—The amount of air inhaled per minute.

mitosis—The continuous event of cell division, consisting of four phases: prophase, metaphase, anaphase, and telophase.

motility—Power to move spontaneously.

myelosuppression—A fall in blood counts.

natural killer cells—Large, granular lymphocyte cells normally present in the body whose normal function is to kill virally infected cells.

necrosis—Death of areas of tissue or bone.

neoplasm—A new and abnormal formation of tissue that performs no physiologic function.

nephrotoxicity—Pertains to toxicity to the kidneys.

neutropenia—Abnormally small number of neutrophil cells in the blood.

nucleus—The vital body in the protoplasm of a cell; the essential agent in growth, metabolism, reproduction, and transmission of characteristics of a cell.

oxidative system—Metabolic system that utilizes oxygen to produce ATP.

oximetry—The determination of the amount of oxygen in the blood.

perfusion—The passing of a fluid through spaces.

pericarditis—Inflammation of the membranous sac enclosing the heart.

peripheral blood flow—Blood flow directed to the periphery of the body, for example to the skeletal muscles.

phagocytes—Cells that have the ability to ingest and destroy particulate substances such as bacteria, cell debris, etc.

platelet-derived growth factor—A growth factor secreted by platelets that can cause vascular cells to multiply and grow.

principle of overload—Principle stating that the physiological system must be stressed beyond normal to stimulate improvement.

principle of progression—Principle stating that the volume of work must be progressively increased to provide an overload to stimulate further improvements.

principle of specificity—Principle stating that physiological and metabolic responses and adaptations to exercise training are specific to the type of exercise and the muscle groups involved.

promotion—The second stage in carcinogenesis, in which the cell begins uncontrolled division and tumor development.

proprioceptive (proprioception)—Referring to the awareness of posture, movement, and changes in equilibrium and the knowledge of position, weight, and resistance of objects in relation to the body.

receptors—Molecular group in cells that have a special affinity for toxins, stimuli, etc.

reliability—The extent to which test results are consistent and reproducible.

sarcomas—A cancer of supporting or connective tissue such as cartilage, bone, muscle, or fat.

senescence—The process of growing old.

serous membrane—A membrane lining a serous (containing serum or a serumlike substance) cavity (e.g., pericardial cavity).

staging—An organized process of determining how far a cancer has spread.

steatosis—Fatty degeneration.

stenosis—Constriction or narrowing of a passage or orifice.

stroke volume—The amount of blood pumped out of the heart per contraction.

submaximal exercise level—Moderate level of exercise, between 120 bpm and 150 bpm.

syngeneic—Referring to tissue obtained from individuals with the same genetic makeup.

synthesis—The process involved in the formation of a complex substance from simpler elements or compounds, such as the synthesis of proteins from amino acids.

tachypnea—Abnormal rapidity of respiration.

T cells—A type of lymphocytes involved in the immune response.

terminal differentiation—Process in which a cell becomes sterile and is unable to divide, thus losing functional properties.

thrombocytopenia—Condition in which numbers of circulating mature platelets are decreased.

thrombopenia—Small number of blood platelets.

tidal volume—The volume of air inspired or expired per breath.

tumor—A spontaneous new growth of tissue forming an abnormal mass.

tumorogenesis—The development of tumors.

validity—The extent to which a test actually measures what it claims to measure.

ventilation—The amount of air inhaled per day. Ventilation rate is the amount of air breathed in 1 min.

Index

Note: The italicized *f* and *t* following page numbers refer to figures and tables, respectively.

About the Authors

Carole M. Schneider, PhD, is a cofounder and director of the Rocky Mountain Cancer Rehabilitation Institute, which helps cancer survivors regain their quality of life through exercise interventions. A cancer survivor, Dr. Schneider is a professor at the University of Northern Colorado. Her background in exercise and cancer recovery includes conducting clinical and basic research and the development of a certificate in cancer rehabilitation. Dr. Schneider is a fellow of the American College of Sports Medicine (ACSM) and a member of the Colorado Cancer Coalition, and she is the recipient of numerous grants from the Susan G. Komen Breast Cancer Foundation. She earned a PhD in exercise physiology from the University of Minnesota.

Carolyn A. "Cad" Dennehy, PhD, is professor emeritus at the University of Northern Colorado (UNC). She cofounded the Rocky Mountain Cancer Rehabilitation Institute at UNC and served as director of programs and education from 1996 to 2002. At present, she is founder and president of Second Wind Cancer Rehabilitation, a privately owned company that provides symptom management services to cancer patients. Dr. Dennehy has published books and articles and authored numerous successful grants. She continues to conduct research on the benefits of exercise for cancer patients both during and after treatment. She is a member of the American College of Sports Medicine and the Colorado Cancer Coalition. Dr. Dennehy received her PhD from Texas Woman's University in exercise physiology. She resides in Estes Park, Colorado.

Susan D. Carter, MD, is a cofounder and medical director of the Rocky Mountain Cancer Rehabilitation Institute. She is involved in all aspects of the Institute, providing leadership and professional expertise to students and faculty involved in clinical practice and research. Dr. Carter is a gynecological surgeon and operates a private practice in Greeley, Colorado. She is a fellow of the American College of Surgery and the American College of Obstetrics and Gynecology. She is an active member of the Cancer Committee at North Colorado Medical Center, director of the Regional Breast Cancer Center of Northern Colorado, and a member of the Colorado Cancer Coalition. Dr. Carter is involved in numerous American Cancer Society and Susan G. Komen Breast Cancer Foundation projects. She is a physician at the North Colorado Sportsmedicine Clinic and is one of the project directors for a breast cancer genetic research project. Dr. Carter received her medical degree from the University of Texas Medical Branch at Galveston. She completed an internship at Hermann Hospital/M.D. Anderson in Houston, Texas.

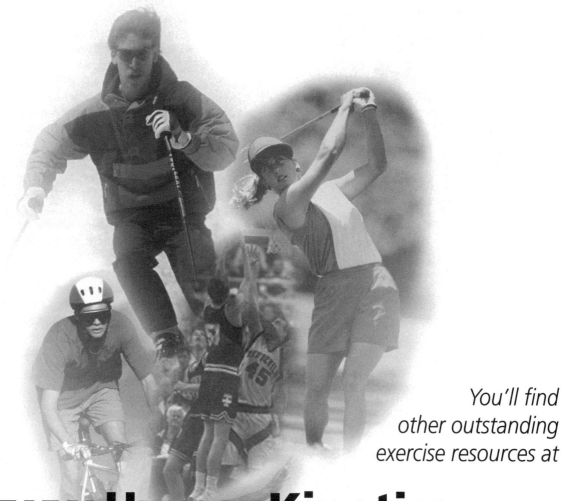